Crispîn Makumbelo El'Shie
Sébastien Luyindula Ndiku
Félicien Lukoki Luyeye

Contribution to Urban Ecology

Crispîn Makumbelo El'Shie
Sébastien Luyindula Ndiku
Félicien Lukoki Luyeye

Contribution to Urban Ecology

Ecodevelopment, Food security, Urban agriculture 1990 - 2000 vers 2025 - 2035 Kinshasa - DR Congo

ScienciaScripts

Imprint

Any brand names and product names mentioned in this book are subject to trademark, brand or patent protection and are trademarks or registered trademarks of their respective holders. The use of brand names, product names, common names, trade names, product descriptions etc. even without a particular marking in this work is in no way to be construed to mean that such names may be regarded as unrestricted in respect of trademark and brand protection legislation and could thus be used by anyone.

Cover image: www.ingimage.com

This book is a translation from the original published under ISBN 978-620-6-70469-0.

Publisher:
Sciencia Scripts
is a trademark of
Dodo Books Indian Ocean Ltd. and OmniScriptum S.R.L publishing group

120 High Road, East Finchley, London, N2 9ED, United Kingdom
Str. Armeneasca 28/1, office 1, Chisinau MD-2012, Republic of Moldova, Europe
Printed at: see last page
ISBN: 978-620-8-10170-1

Contents

This book is dedicated to **Professor Emeritus Jacques Paulus SJ**. for his commitment to Ecology and Urban Agriculture in Kinshasa through :
- his teaching and research in the 3eme Cycle Programme in Environmental Management at the Biology Department, the UNESCO Chair and the Environmental Sciences Department at the Faculty of Sciences of the University of Kinshasa;
- its service to households that had at least one malnourished stage III child through the Jardins et Elevages de Parcelles (JEEP) project;
- who left us almost a decade ago.

We also said this to **Marie Makumbelo Nkiri " Santa"** dëcëdëe just after presenting the first set of results from this study to a jury meeting at the University of Kinshasa - Kinshasa (Rëpublique Dëmocratique du Congo).

Resume

A study of the population's reaction to the effects of urban sprawl was carried out in Kinshasa (DR Congo). The households of Kinshasa from 1990-2000, shaken by a multifaceted crisis caused by the effects of the destruction of the economy due to non-compliance with the resolutions of the Sovereign National Conference, the war of liberation and the breakdown of the capital's food supply routes, included urban agricultural production among the solutions to combat the serious health problems of their members. A survey of a sample of 201 plots of land in the Commune of Limete, various urban farmers and agricultural support structures, complemented by a nutritional survey of children aged 0-5 and their mothers, showed that with 19 species of vegetable and 47.000 trees of 18 fruit species, this population produced 33.27 tonnes of vegetables and 4,087.00 tonnes of fruit per year. With a production of 36.67 tonnes, *Mangifera indica*, *Persea americana*, *Elaeis guineensis*, *Carica papaya*, *Dacryodes edulis* and *Musa paradisiaca* (and citrus fruits) contributed 10.9 g, 6.1 g, 4.5 g, 4.4 g, 1.8 g and 0.6 g of fruit per person per day respectively. This equates to an average availability per person per day of 4.54 Kcal, 5.82 Kcal, 14.58 Kcal, 1.04 Kcal, 2.37 Kcal and 0.34 Kcal respectively. In view of the urban sprawl, the population explosion, the fragmentation of housing plots, the subdivision of market garden sites and the conflicts caused by competition for farmland on the outskirts of the city, the households of Kinshasa 2025-2035 need to envisage urban agriculture that is adapted to the cities that are being built. Soilless farming will have to be integrated to boost food availability.

Key words: urban ecology, urban unplugging, Kinshasa, urban agriculture.

INTRODUCTION

01. State of research and problems

Ecology is the study of the interrelationships between living beings and their environment. It places this subject at the top of the organisational scale of living beings, where it studies biological processes at the highest level.

I decosvsteme (forest, grassland or urban for terrestrial ecosvstemes) is the dogma of the environment in the same way that DNA is the dogma of genetics. In a balanced ecosystem, there is a set of reciprocal exchanges between living beings and their habitat. Otherwise, we would say, paraphrasing Dajoz (2000) who takes up Clement's (1916) theory, that the biotope exerts an 'action' influence on the biocenosis and the biocenosis reacts by exerting a 'reaction' influence on its biotope (habitat). On a deeper level, we would conclude that the influences exerted by a habitat create an action on the living beings that populate it. In return, the latter respond by reacting in order to cope with them. In simple terms, we would say that living beings are always seeking stability in order to survive in a constantly changing environment. Man is not immune to this struggle.

This is why, faced with the difficulty of answering the question: "how can we feed everyone, given the number of people living in more than one city, especially in terms of calories, vitamins, mineral salts and other nutrients required by the human organism"? Ramade adds that "of all the serious environmental problems that characterise the present time, that of food availability is such as to worry the least pessimistic of ecologists" (Ramade, 1982).

The problem of food availability cannot be solved by slogans, and even less by continuous food imports, especially when you consider the difficulties of storage, preservation and the perishable nature of foods such as fish, vegetables, fruit and meat.

Food availability is only one component of food security. Food security also includes economic and physical access, utilization and stability, and food safety and quality (World Bank, 2023).

According to the political dictionary, food security is the situation that guarantees all human beings, at all times, the physical, social and economic possibility of obtaining sufficient, healthy and nutritious food. It must be sufficient to ensure a healthy and active life, taking into account eating habits (https//www.touphie.org.).

Food security exists when all people, at all times, have the physical, social and economic access to sufficient, safe and nutritious food to meet their dietary needs and food preferences for an active and healthy life (Committee on World Food Security, 2012).

The Rome Declaration on Food Security recently stated: We proclaim our political will and our common and national commitment to achieve food security for all and to make an unremitting effort to eradicate hunger in all countries and immediately reduce by half the number of undernourished people by 2015 at the latest.

Food safety is traditionally considered to have four dimensions or pillars:

1. accds (the ability to produce one's own food and therefore have the means to do so, or the ability to buy one's own food and therefore have sufficient purchasing power to do so);
2. availability (sufficient quantities of food, whether from domestic production, imports or aid) ;
3. quality (of food and diets from the nutritional, health and social-cultural points of view);
4. stability (of access capacities and therefore of prices and purchasing power, availability and quality of food and diets).

From this perspective, food security has a more technical dimension. This distinguishes it from the concepts of food self-sufficiency and the right to food, which have more political or legal dimensions.

Commercial food imports are de facto subsidised (Gossens et al., 1994) but in no way guarantee that all sections of the population will have access to good quality food in sufficient quantity and at regular intervals. We must never lose sight of the fact that food aid is a weapon of dependence and instability for local production structures if we are not careful. In fact, obtaining cereal surpluses is inevitably a weapon. Food is ammunition (Sachs et al., 1981). Only local production can fully guarantee food security.

While it is true that some of the vegetables, fruit, rice, fish, meat, mushrooms and other foods consumed by households in Kinshasa come from outside the DR Congo, it is also true that some are produced inside the country and outside the city. A third part is the product of urban agriculture in the city itself. It is all of this production that consolidates the food security of urban households.

This agriculture plays an important role in supplying food to the capital (Kinshasa). As a result, these urban farmers contribute to the availability of food, and even more to the food security of the city's households. In the sense that they ensure that farm households have stable and permanent access to locally-produced food.

Generally speaking, the best-known factors underlying food safety are essentially :

1° the political and socio-economic context ;

2° the degree of functionality of the economy ;

3° care practices in health care institutions, and

4° health and hygiene in the home.

02. Research objectives

The general aim of this work is to answer the question: "How can we feed everyone in Kinshasa, rather than how can we teach everyone about agriculture and livestock farming? This objective is broken down into two specific objectives which are studied:

- the contribution of urban agriculture to household food security. As food security is very closely linked to the health of the population, the work will also look at the nutritional status of children of preschool age to gain an idea of this; - the ecodevelopmental possibilities for improving this agriculture for the good health of all.

03. Working hypothesis

In response to the specific research problems that this text no longer addresses, the fundamental assumption is to "make food accessible to the entire population of this capital". This statement will be dissected as follows:

- the lack of food available in households is contributing to the deterioration in the population's health;

- Some of the food (vegetables, mushrooms, fruit, seeds, fish, meat) consumed in Kinshasa is produced (or picked) locally;

- the majority of those who practise urban agriculture show very little interest in taking concerted and sustained action among themselves, even though they are often financially deprived and inadequately supervised households;

- In general, they practise the usual techniques, without making any effort to improve them;

- Coordination of all the structures involved in urban agriculture would be an asset for the eco-developmental improvement of these activities, which could make a lasting contribution to household food security in the city of Kinshasa;

- Kinshasa's urban sprawl is prompting a new scenario for agriculture to boost household food security in Kinshasa - the case of Limete (one of Kinshasa's 24 communes)

04. Framing the study

The research is rooted in urban ecology and ecodevelopment. Urban Agriculture and EthnoSciences are merely tools for appreciating this environment-development interdependence for the human-urban habitat interrelationship in an urban ecosystem.

This is how the research was carried out:

- the survey to identify the existing opportunities for urban agriculture, the various players and partners, the objectives set, the constraints encountered and the production collected, and to identify the expectations of parents for the good nutrition of their households;

- Weighing, measuring the height of children aged 1 to 5 and collecting statistics on children born with low birthweight to assess the health of the population.

This book is more about trying to feed everyone well than about trying to teach the population about agriculture or animal husbandry as described in the preceding lines.

This is a concern which can only find an appropriate solution today if it is unanimously accepted that the interdependence between the environment and development leads to recognition of the need to associate these two fields in all reflections concerning the search for solutions to the problems of human societies (Matuka, 1999) and also that ecology and the economy are indeed closely linked. Poorly designed development can degrade the resources on which it is based. Similarly, environmental degradation can lead to the failure of economic development, whatever the cost. Hence the need for Ecodevelopment. In order to help solve the problems of malnutrition in Kinshasa, ecodevelopment calls for better management of the resources of the urban environment in general and of each plot of land in particular (Paulus et al., 1989). This vision of development calls for the use of ecological and economic agriculture (Mercier, 1980), which aims to feed people using environmental resources while respecting the balance of the ëcosystëmes. This perspective proposes the implementation of alternative methods that ensure food security for populations and reduce endemic malnutrition. It is also a production method that takes into account the traditional knowledge of farmers and integrates scientific progress from all disciplines. It is a method that makes it possible to combat child malnutrition at community level (Brown & Brown, 1977).

05. Interest and choice of subject

For Science, this work brings together Ecology (urban synecology), Ecodevelopment, Food Security and Agriculture in the search for solutions to the problems of the crisis of unbalanced societies.

In practice, he describes the situation of a town already destroyed by the reaction of the population (fighters) to the refusal to respect the Sovereign National Conference and then shaken by the war of liberation led by the AFDL and others (19961997). This destruction created a multifaceted crisis, accentuated by the destruction of the roads linking the interior of the country (the centre of food production) and its capital, Kinshasa (the centre of consumption). Households, deprived of all availability and accessibility to food, are committed to exploiting the empty spaces of housing plots, peripheral gardens and other sites in order to produce, at low cost, basic foodstuffs to feed all their members. It is this reaction of city dwellers in a crisis situation that is referred to as urban agriculture. It is this picture of Kinshasa in the 1990-2000 decade that this book paints. It then looks ahead to the 2025-2035 decade, when other similar events, such as the population explosion, the fragmentation of housing plots, the subdivision of former market-garden production sites, and conflicts between tribes over the occupation of land in the city's rural areas, will prompt a new search for solutions to feed this ever-growing population in a city that has continued to grow out of the ground over the decades.

Finally, it reconciles data from more than two dëcennies, for a literature capable of inspiring households, facilitators and decision-makers to more humanistic reactions in a world whose environmental balance is still under threat.

06. Defining the subject

This study will be limited to analysing the level of food security (including nutritional status) and the state of urban agriculture (maintenance of production sites and production) in the Commune of Limete.

The basic data for this study was taken from the years 1990 - 2000. The surveys were conducted in 1995 and 1996. Observations of the results have continued to the present day, with a number of publications in the meantime.

LITERATURE REVIEW AND METHODOLOGICAL APPROACH

This part deals with: - the conceptualisation of the study, - the description of the study environment and - the methodological approach.

Chapter 1: SETTING UP THE STUDY

Five basic concepts form the pillars of this study. These are: urban eeology, the systemic approach, ecodevelopment, urban agriculture and food security. However, as the last of these conceals the idea of a population's food situation, which can be determined by the level of children's nutritional status, it is easyë to also describe the notion of protein-calorie malnutrition in preschool-age children.

1.1. Urban ecology

This sub-discipline of ecology has its origins in the gardens of Babylon. The Chicago School is recognised as the founder of the first urban movement. With an approach less linked to scientific ecology and more closely related to sociology in the 20th century (^{eme}), this school thought about the interdependence between city dwellers and their environment.

In the years 1990-2000, with its notion of ecological footprint, the school described the city as a more or less natural area. Later, the city was seen as a source and sink of flows and energies, with complex direct and indirect impacts on biodiversity, the biosphere and the climate, and with special relationships between city dwellers. The urban community is seen as both a spatial model and a moral framework.

The trend in this sub-discipline can be seen in the work of the Rio de Janeiro Conference (June 1992), with terms such as "city renewed upon itself", "construction seeking to repay its ecological debt and reduce its ecological footprint".

Urban ecology is the study of all environmental problems in the urban environment. Its aim is to integrate these issues into local policies in order to limit environmental impact and improve the quality of life for local residents.

Strictly speaking, urban ecology is a field of ecology that focuses on the study of the city as an ecosystem. Today, by popularisation and with the aim of raising awareness of environmental issues, it can encompass the consideration of all environmental issues concerning the urban and periurban environment. It aims to articulate these issues by integrating them into local policies to limit or repair environmental impacts.

Urban ecology postulates an interdependence between city dwellers and their urban environment, which the notion of the ecological footprint will extend to the planet in the years 1990-2000.

Its challenge is to preserve a place for biodiversity in the city. This means combating ecological fragmentation by maintaining or restoring biodiversity continuity in towns and cities fragmented by roads and/or buildings, for example.

Cities are not always biological deserts devoid of interest for plant biodiversity in particular. Several sacred places and corners of more than one city are veritable relics of the prehistoric forest (www.techno-Science.net).

This biodiversity is often тепасёе and even totally dëtruite by the urban dëbranchement brought about by modernisation and the culture of concreting everything (Faburel, 2023).

It is this last aspect of urban ecology that interests this book. It is well known that when the quantity of a resource is limited, competition arises. This competition quickly moves from indirect competition to direct competition when the quantity is further reduced.

1.2. Systemic approach

A system is a set of dynamically interacting elements that are organised to achieve a goal. Its essential characteristics are :

1) causality (linear): this is determinism. Everything is predetermined; every fact has a cause; under the same conditions, the same causes produce the same effects;

2) interaction (between the elements of a system): this includes feedback or retroaction which modifies the behaviour or material of the components of the system. It is at this level that we have the retro-positive loop (when it has an amplifying effect on all movements) or the retro-negative loop (when it has a regulating effect on maintaining these movements);

3) totality: a svstem is a whole ;

4) organisation: this is the central concept of a system, the arrangement between elements or individuals with a view to producing a new unit;

5) complexity: the complexity of the number of elements and the complexity of the relationships linking these elements (De Rosnav, Sd quoted by Makumbelo et al., 2023).

The systemic (or holistic, synthetical) approach consists of considering a complex system in terms of its emergent characteristics linked to its totality, properties which cannot be reduced to the sum of its parts. It is the science of describing and explaining the diversity of living forms. It is also an approach that emphasises the links and interactions between parts (Lacoste and Salanon, 1999; Schwarz, 1997 cited by Gobat et al., 2003; Giordan and Saltet, 2011 cited by Makumbelo et al., 2023).

The systems approach is based on its own cycle: model, scenario and act.

The effectiveness of the systemic approach lies in the fact that it can consider two, three or four variables at a time when studying the effects of the interactions of the system instead of only one, sometimes, which the analytical approach considers in order to grasp the nature of the interactions (De Rosnav, Sd).

Urban agriculture is a case of the agricultural production system, - for which, urban agriculture can only be mastered if the methods of the analytical approach are supplemented by those of the synthetical approach. The synthetical approach enables the structural and functional aspects of this type of agriculture to be highlighted, its boundaries to be defined, all the sub-systems and existing interactions to be identified, inputs and outputs to be analysed, and the various relationships to be analysed, before proposing a model and establishing its simulation in the field.

1.3. Ecodevelopment

The concept Ecodëveloppement composë d'Eco ^conomie, ëcologie) et dëveloppement can only be understood if it is p l acë in its historical context.

The concept of the environment emerged in the 1970s. For the concept of development, the 1961-1970 decade was marked by the launch of the first period of development, following the Cairo Conference. This decade was marked by Western assistance to poor nations. At the end of the second decade 19711980, assistance was replaced by a more mercantile approach. This decade saw the West export its industrial growth model to developing countries. Towards the end of the 1970s, there was a period of reflection on the involution of ideas in the field of the environment - following the Stockholm Conference (United Nations Conference on the Environment, 1972 ; www.uni.org) and on development. The new ideas put forward by both UNESCO and the World Bank have given rise to new concepts: integrated development, ecodevelopment, sustainable development, human development and ecologically viable development.

The ëcodëveloppement implies that the concenrees populations organize and s^duce themselves to better apprehend the spëcific possibilities of their ecosysteme and to develop them with the help of the appropriate techniques spëcially congues to this adapted end or, in some cases, deliberately imitated. Relying on one's own strengths (self-reliance) does not mean isolating oneself in a more or less impossible autarky, but rather deciding on an autonomous fagon where original research must be carried out and where it is appropriate to borrow from the outside when it is necessary to renew with tradition and when, on the contrary, a break is necessary. This vision of development favours ecological development styles. The most striking features of the ëcodëveloppement can be summed up in several points, including those of interest to this work:

1. In each ë^дю! the focus is on developing its specific resources to meet the population's basic food needs. These needs ëbeing dëfined in a realistic and autonomous manner so as to avoid the harmful demonstration effects of the consumption style of the rich countries.

2. Since man is our most precious resource, 1 ëcodëdevelopment must above all contribute to its realisation: employment, security, quality of human relations, respect for the diversity^ of others or, if we prefer, the establishment of a social ëcosystem that is judgedë to be satisfactory.

3. The identification, development and management of natural resources are carried out in a perspective of diachronic solidarity with future generations.

4. The negative impact of human activity on the environment is reduced by using processes and forms of production organisation that make it possible to take advantage of all available resources and use waste for productive purposes.

5. The ëcodëveloppement implies a particular technological style. Eco-techniques exist and can be implemented for food production among others by new imaginative ways of industrialising renewable resources. The development of eco-techniques is destined to take a very important place in ëcodëvelopment strategies for the good reason that the capitalisation of various objectives - economic, social, ecological - can be done appropriately at this level.... . But it would be wrong simply to equate ëcodëdevelopment with a technological style. It also implies methods of social organisation and a new system of education.

6. The institutional framework for ecodevelopment cannot be defined once and for all without regard for the specificity of each case, following three basic principles: - horizontal administration, effective participation of the populations concerned in the implementation of ecodevelopment strategies for the definition and harmonisation of concrete needs, the identification of productive potential and the organisation of the collective effect for its development (Touraille, 1977). It should also be noted that there can be no ecodevelopment without environmental education (Sachs et al., 1981).

At present, we are more inclined to talk about sustainable development. But the future will clearly have to focus on the following aspects:

1. the practical application, in favour of development, of the panoply of theoretical resources developed over the last three decades, in particular Agenda 21. In other words, we need to take action, but thoughtful action.

2. 1. the adoption and implementation of a moral code of action (a kind of ethical revolution) to protect the global environment against irreversible damage, and to introduce greater equity into development.

1.4. Food safety
1.4.1. Protein-calorie malnutrition
Protein-calorie malnutrition is one of the nutritional problems affecting individuals or groups of individuals (or communities). The indicators height-for-age, weight-for-height, weight-for-age, gill perimeter, birth weight and skinfolds are used to determine the incidence of protein-calorie malnutrition in children of preschool age.

Each of these indicators determines a type of malnutrition. For example, the height-for-age (HFA) ratio of children aged 0-5 years indicates chronic malnutrition and determines the child's nutritional history. It is, in fact, the result of the individual's adaptation to lasting poor social, health and nutritional conditions.

It is a valid indicator of food consumption and the health status of the population in the medium term.

Malnutrition " ;iiguc" " reflects the individual's short-term maladjustment to the
poor nutritional conditions. It is measured by the child's weight/height (W/H) ratio.

Global malnutrition or "underweight" is the combined result of the effects of chronic malnutrition and "iiguc" malnutrition. It is measured by the weight/age (W/A) ratio (CEPLANUT / FAO, 1994).

Malnutrition can take the form of weight loss or a reduction in brachial perimeter. For this reason, measurement of the brachial perimeter is another tool for screening for malnutrition in children aged 0-5 years (CEPLANUT, 1996). It has been found that during the first year of life, the circumference of the upper arm of a healthy child increases rapidly. Then, from the age of 1 to 5, it remains more or less the same diameter. But the arm of the malnourished child always remains thin. So between the ages of 1 and 5, if the circumference of the upper arm reaches 13 cm, the child is well nourished. If not, the child is malnourished (CEPLANUT, 1996). The brachial perimeter also determines acute malnutrition. For the WHO, low birth weight is an important indicator of the mother's health. It reflects the expression of weight, health status (gënito-urinary disease, malaria), diet and nutritional deficiencies (nutritional anaemia, avitaminosis A, maternal iodine deficiency) (Toko et al., 1980 cited by CEPLANUT, 1994). In DR Congo, there are virtually no older publications on full-term births of underweight children (CEPLANUT / FAO, 1994). In addition, Body Mass Indices (BMI) can be used to determine the nutritional status of adults. This is the case of the Body Max Index (BMI) or Quetel indicator.

1.4.2. Food safety proper
Food security means either being able to produce and give everyone access to sufficient, healthy and balanced food on a permanent basis, or having the money to access it easily through the market. It is not the fact of having money that is important, but it is above all the fact of having a stable supply of food and easy access to sufficient, healthy and balanced food that matters. This implies that access to sufficient, good-quality food should be available to the entire population without exception. This leads us to believe that we can only talk about food security if everyone, including the most disadvantaged, has access to this food. This is because its ultimate objective is to ensure that all human beings have material and economic access at all times to the basic foodstuffs they need (Azoulay, 1998).

The most important factors in determining the food security of a household are the availability of a sufficient quantity of quality food, its stability and its accessibility to all members of the household.

Availability is made up of internal protection, imports, excessive exports and aid.

On more than one occasion, as in its Plan Directeur du Dëveloppement Agricole et Rural (Master Plan for Agricultural and Rural Development), the Congolese government has determined to ensure access to sufficient, balanced and regular food for all sections of the population, with particular emphasis on the most vulnerable groups (Ministere de l'Agriculture, Animation rurale et Developpement communautaire, 1991).

The availability and accessibility of good food for the poorest people on a permanent basis, in a normal food production and distribution system, reassures us that the entire population can feed itself well and can ensure a certain degree of food security. In other words, the food security of a household does not simply mean the accessibility of food, but it also means that the food available is qualitatively good and quantitatively sufficient and accessible on a stable and sustainable basis to all members of the household.

Food security differs from food self-sufficiency in that for the latter, it is the possession of a sufficient quantity of food that takes precedence over accessibility and stability. This explains why, in more than one case, some members of the community, especially the poorest, suffer from malnutrition and starve to death next to the large stores of food on the spot. At the round table held in Kinshasa from 4 to 8 March 1991, it was recommended that the aim should no longer be to achieve food self-sufficiency at any cost, but rather to create conditions that would guarantee access to sufficient, balanced and regular food for all sections of the population (Goossens et al., 1994). When food security is assessed using the food balance sheet method, it refers to the population's food supply and demand.

1.5. Urban agriculture

1.5.1. Agriculture

The concept of agriculture is derived from the words "ager" (field, earth, soil) and "colere" (to cultivate, plough). Agriculture is therefore the action of cultivating and ploughing a field. So when we talk about agriculture to the uninitiated, it simply means growing a plant or raising an animal. In reality, however, it is the art of obtaining from the soil, while maintaining its fertility, the maximum amount of useful products from either the plant or animal kingdom. The term agriculture refers above all to techniques for producing plants or animals.

For scientists, this term brings to mind the concept of agronomy, which is the study of the laws that govern agriculture and which makes up a group of sciences called Agronomic Sciences. These sciences are essentially composed of plant science and animal science.

The farmer's work involves plants, fields, trees, animals, poisons and bees. His aim is to enable the reproduction of living species to ensure the survival of human beings. Thus, this creative activity of man in nature is called agriculture, mycoculture, arboriculture, forestry, fish farming and beekeeping when it is applied respectively to the soil and the plant, mushrooms, trees, the forest, fish and bees. Myciculture or mushroom cultivation is not to be confused with mycoculture, which is a laboratory cultivation technique used in biology (medical mycology) for mycetes of medical or veterinary interest.

In this book, as it is only about its practice and not its theory, agriculture is nothing other than man's creative activity in nature, applied to the soil and plants (rice and various vegetables), mushrooms, trees (those whose fruit is consumed by man), poisons and domestic animals. In other words, this work focuses on fruit growing, livestock farming, market gardening, mycoculture and rice growing.

The term "market gardening" here includes both market garden crops grown on large production sites and kitchen garden crops grown on plots, as long as part of the production of both is always sold.

In botany, the term "tegume" refers to the fruit of teguminous plants (pods or dried fruit that open when mature with two longitudinal slits): *Fabaceae* (*Faboideae, Mimosiodeae* and *Celsalpiniodeae*) whose bacterial nodosites are placedëes tangentially on roots), in this work, it designates any herbaceous, monthly or annual or even perennial plant of which one of the parts (stem, leaf, pod, flower) is consumed without indispensable industrial transformation.

1.5.2. Urban agriculture

1.5.2.1.　　Historical development

At the beginning of the XIXeme Century, the ships calling at the African coasts began to create a market around the ports. Later, missionaries planted gardens for their own consumption. Before independence, market gardening centres sprang up near military camps. From the end of the 1960s, African towns grew in importance, with the rural exodus and demographic pressure encouraging the development of urban centres. Around the capitals, market gardens spread out. A mixture of local and exotic vegetables is grown. The growth of cities and the strong demand for vegetables that goes with it are the driving force behind the development of market gardening. The 1973 drought in West Africa triggered a major boom in market gardening. It appeared to be an interesting and remunerative alternative, at a time when production systems were highly disorganised. It has also reduced the number of non-governmental organisations (NGOs), all of which have set out to support market gardening projects. Many NGOs are thus created, sometimes somewhat artificially, and above all without much coordination (Autissier, 1994). Kinshasa is not immune to this movement. Although they no longer have well-developed market garden belts, urban farmers produce food wherever they can.

Since the 1990s, urban agriculture has been a key issue in sustainable development, urban planning and the fight against food insecurity, not only in the countries of the "South", but also in the rest of the world.

According to Boulianne (2016), its importance in supplying people living in cities is now inescapable in many cases. This importance is justified according to the FAO (2014) by the fact that 1 urban agriculture is, ëeconomically, a source of food and employment. In addition, ëecologically, it makes cities more "viable" because they are more resilient in the face of climate change. But also socially, it offers opportunities to Edification of human relations through its inputs and outputs.

1.5.2.2.　　The scope of urban agriculture

Urban agriculture and by extension, urban and рёнш^^ is a form of agriculture that ë emerges from agricultural practices rëalisëes in cities. Currently, on a planetary scale, we are witnessing a growing interest from the various actors in sociëtë for urban agriculture projects as a vector of ecological transition (sustainable food, social link and wellbeing of populations, ë environmental education, ecosystems, landscapes), community (participatory projects), etc.

The subject of several ëtudes, urban agriculture is agriculture practised in an urban environment. This type of agriculture allows city dwellers to eat vegetables, meat and starch produced by their own efforts (Paulus et al., 1989).

By producing their own food, city dwellers will be able to improve their nutrition and, in turn, ensure their good health. They will be able to lighten their family budgets, a large part of which is usually spent on household food supplies.

The sale of surplus production can increase household income. Food production by city dwellers also helps to create jobs in the city (producer-seller, buyer-seller, processors, transporters). This role has already been mentioned in the preceding lines, where the writings of Boulianne (2016) and the FAO (2014) are cited. This is what makes it difficult to find open land in the urban areas of north-west Cameroon. This uncultivated land has been ploughed; playgrounds and squares have been put under cultivation (Ashah, 1997).

Urban agriculture that is protected, stimulated and supervised by the public authorities and others can become a real link in the development chain. Development organisations recognise that market gardening is highly profitable per unit area, has a high nutritional value and is easy to integrate into traditional production systems.

To answer the question "What is the advantage of diversifying market gardening in tropical countries? Messiaen points out that market gardening makes it possible to reduce imports of fresh and preserved vegetables from temperate countries, and also to diversify exports (Messiaen, 1974). Development organisations almost always use these arguments to justify their interventions (Autissier, 1994).

Chapter III. STUDY ENVIRONMENT AND SITUATION OF FOOD SECURITY AND URBAN AGRICULTURE

2.1. Study environment : Commune of Limete / Kinshasa

2.1.1. Geographical, geographical, climatic and hydrographic situation

One of the 24 communes of the capital of Kinshasa, the Commune of Limete covers an area of 27.1 km^2 (INS, 1984). According to en.m.wikipedia.org, the surface area of the Commune of Limete in 2023 is estimated at 67.60 km^2 .

The boundaries of the municipality and its various districts are shown in Figure 1.

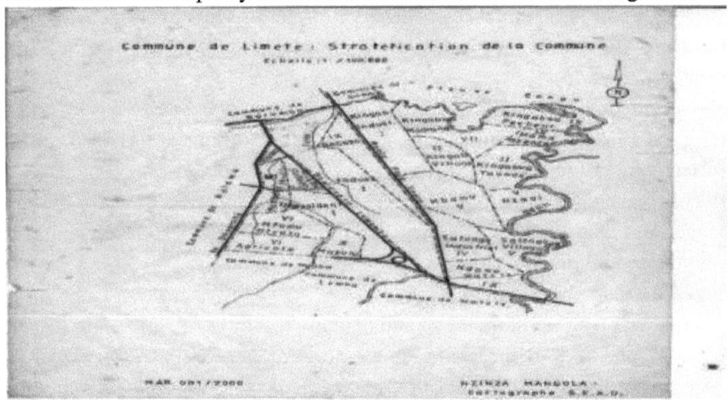

Figure 1: Map of the Commune of Limete.

This figure shows that three major roads cross the commune. These are Boulevard Lumumba, Avenue Universite and Route Poids Lourd. A ëchangeur is located on Boulevard Lumumba not far from the Foire de Kinshasa facilities. The map shows the different districts of this Commune.

As with the city of Kinshasa as a whole, the Commune of Limete is located at latitudes 4°18° and 4°25° South and longitudes 15°15° and 15°22° East.

Its soil belongs to the group of sandy tropical soils rich in iron and alumina and subject to the action of a hot, humid climate. According to the system proposed in 1979, the climate of the city of Kinshasa belongs to the AW4 type, i.e. a humid climate with a dry period of 4 months

14

from May to August or June to September and a rainy period of 8 months, from October to April or May depending on the year. The average tempërature hovers around 29°C, with March and April the hottest months and July the coolest. It has a hot and humid tropical climate with a regular rainfall pattern averaging around 1,500 mm (Agence Nationale de Meteorologie et Teledetection par Satellite-Station de Binza, 2015).

The N'djili river separates the Commune of Limete from that of Masina and flows into the Congo River. As it flows through, it creates a valley covering more than 500 hectares in its lower reaches.

2.1.2. History of the Commune

The history of Limete is linked to that of the city of Kinshasa, in the sense that in 1955, as with all the large conurbations in Africa at the time, Kinshasa was divided into only two districts: that of the whites and that of the indigenes (Houyoux, 1973). The Commune of Limete, like that of Gombe, was one of the residential Communes inhabited only by whites.

Since then, many things have changed. Three important periods mark the division of the Commune. These were the period before the 1974, 1975 and 1976 divisions.

The latest division is that which created the N'danu district by splitting the Salongo and Maman Nzenze entities. Unnamed sources write that this division was mainly influenced by political and land issues.

The Commune has 14 neighbourhoods based on the most recent decoupages (Mfumu mvula, Kingabwa, Mombele, Mososo, Quartier industriel, Salongo, Quartier residentiel, Masiala, Mateba, Mayulu, Agricole, Ndanu, Mbamu, Nzadi, Russell Matou).

2.1.3. Demographic situation

The Commune is inhabited by 128,197 people, including 65,768 men and 62,429 women in 20,832 ordinary households and 34 collective households. The population density is 4,731 inhabitants per km^2 (INS, 1992). It is distributed as shown in Table 1. Gutu kia Zimi (2021) reports that the population of the Commune of Limete was 502,459 in 2020, with a density of 4,370 inhabitants per $km.^2$

Table 1. Rëpartition of the population residing in the Commune of Limete.

Administrative entities by district	Residential population	Number of plots per neighbourhood	Ordinary households	Group households
Mayulu	13.729	1.123	2.251	-
Agricultural	3.992	216	693	-
Mombele	4.367	883	1.425	-
Mateba	2.092	337	327	-
Musoso	5.081	3124.	710	-
Kingabwa	15.941	2.210	2.689	-
N'danu	15.017	1.500	2.530	2
Nzadi	12.436	550	2.422	1
Salongo	8.622	1.094	1.179	2
Mfumu-Nvula	20.083	1.299	3.356	1
Industrial	4.700	1.358	748	13
Masiala	6.209	360	858	1
Residential	3.695	950	443	2

Bobozo	487	-	70	-
Bukaketese	393	-	68	-
Campde Military reg.	236	-	23	8
Funa	2.643	-	359	2
Kwela	343	-	73	-
Victory	697	-	106	-
Yaounde	2.240	-	377	-
Yassa	384	-	77	-
LIMETE	128.197	18.480	20.832	34

Source: INS, 1972.

This population must surely have changed, which is why this work prefers to start recalculating it for updated estimates.

While the average number of people per household in the Commune is not yet known, the average number of people per household in the city as a whole is shown in table 2.

Table 2. Evolution of the average number of people per тёпаде in Kinshasa from 1956 to 1995.

AniK'e	Persons responsible for the study and types of household	Average number of people per mënage	Sources
1956	L. Baeck (out of 46 Congolese ëvoluës).	4,6	Houyoux, 1973
1962	Direction de la Statistique et des Etudes ëconomiques (out of 52 State employees earning between 4,000 and 12,000 francs/month)	-	Houyoux, 1973
1963	P. Caprasse and G. Bernrd (out of 61 monitors)	4,5	Houyoux, 1973
1973	Houyoux (Budgets mënagers: nutritionand way of life of the 1,471 African households that live in Kinshasa	5,9	Houyoux, 1973
1989	Direction desenquetes ëconomic (INS-DëpartementduPlan) (BudgetInquiry mënagersVillede Kinshasa, 1985)	7, 0	INS, 1989
1995	ISAZ (RapportSantë public urban, City of	7,0	ISAZ, 1997

	Kinshasa, 1995)		
2005	Contribution of fruit trees to food sëcuritë in a tropical urban environment (based on a sample of 231 households surveyed)	7,5	Makumbelo et al., 2005

Analysis of this table shows that the average number of people per household rose from 4.6 to 7.5 in 1956 and 2005 respectively.

This average number of people per mënage will surely be even more ëкyë in view of the new developments in this city.

2.1.4. Socio-economic situation

The population of the Commune of Limete engages in urban agriculture, small-scale trade, handicrafts and various services. In Kinshasa, local and imported goods are available, but at unaffordable prices (2 to 3 times the Belgian price). Most of the population shops at markets (ISAZ, 1997). They are normally supplied all year round. Nonetheless, prices for products such as manioc, maize, bananas, rice and others rise from October to April because of the deterioration of roads caused by rainwater, which makes it difficult to transport food products from the Bandundu and Bas-Congo provinces to Kinshasa. However, as early as 1986, it was observed that wage levels remained very low, particularly in the public administration, which played a guiding role for the private sector in this area (Houyoux, 1973).

A recent report states that the salary of a civil servant varies between 0.5 and 3 US dollars per month. The average salary in the private sector at the time rarely exceeded 100 dollars a month (Goossens, 1994).

The percentage change in household expenditure on food in Kinshasa shows a certain upward trend, as shown in Table 3.

Table 3. Trends in the structure of food expenditure per household in Kinshasa.

Years	% of budget spent on food	Studies carried out by
1956	53,0	out of 96 evolved Congolese (Houyoux, 1973)
1962	36,0	out of 52 government employees earning between 4,000 and 12,000 francs/month (Houyoux, 1973)
1963	67,4	on 61 monitors (Houyoux, 1973)
1986	62,0	Goossens (1994)
1993	91,0	CEPLANUT (1996)

This table shows that the percentage of the budget devoted to food per household rose from 53.0% to 91.0% between 1956 and 1993. 62.0% represents almost 2/3 of the money spent (BEAU, 1986).

As for monthly expenditure in 1967, the average expenditure for a household of 5.9 people was 32.83 Zaire (Z) / month (1 Z = 2 US dollars), i.e. 1.41 Z of transfer expenditure and 31.42 Z of consumption expenditure ... Consumer expenditure amounts to 21.19 Z / mënage / month and represents 67.4% of consumer expenditure.

Meat and fish account for 40% of food expenditure, starch and cereals for a quarter, vegetables and legumes for 13.4%... Only households with an income above 60 Z have a satisfactory diet. But even at this level, calorie requirements are not fully covered (19.2%).

The majority of consumption units (71.1%), living in households where monthly expenditure is between Z15 and Z60, can be classified as undernourished (Houyoux, 1973).

A survey carried out in Bumbu and Kingasani in 1996 showed that an average household would spend the equivalent of 3 US dollars a day, or 90 dollars a month, on food (CEPLANUT, 1996). Not everyone can afford this.

A comparison of the quantities consumed per product in 1969, 1973 and 1986 per month and per person shows that the weight of quantities consumed per product is gradually decreasing for protein-rich foods. This already raises fears of malnutrition in the near future.

Table 4 shows the quantities consumed by product from 1959 to 1986. For reasons of space, this table will be ⌃⌃ё year a and b.

Table 4a shows the quantities consumed by product from 1959 to 1986.

Table 4a. Quantities consumed by product in 1959, 1975 and 1986 (per month and per person in Kg).

Products	1969	1975	1976
Cassava	6,120	5,379	5,599
Bread	1,770	1,172	1,575
Leaves of manioc	1,590	1,333	1,296
Banana	0,923	0,423	0,573
Rice	0,607	0,738	1,067
Fresh sea fish	0,516	0,533	0,630
Dirty fish	0,433	0,238	0,086
Beans	0,389	0,369	0,331
Boneless beef	0,361	0,204	0,284
Tomatoes	0,359	0,215	0,189
Sugar	0,350	0,467	0,426
Total	13,418	11,071	11,256

Source: BEAU, 1986.

These 11 products made up Kinshasa's food basket in 1969. The situation in 1986 was fairly similar to that in 1976. There was a tendency to consume more cassava, including cassava leaves, and less sate fish, beans, rice and fresh fish. Bread consumption was lower in 1975. Consumption in 1976 did not reach the 1969 level (BEAU, 1986).

The ётements gathered in this point explain the need to encourage each тёпаде to organize at least one agricultural activity and those who have l agriculture in their charge to fight for the improvement of the conditions of food production by mёnages to count the range of food obtained by the market in order to ensure food security for all.

2.1.5. Urban agriculture

ln Kinshasa, this type of farming is practised on residential plots (housing plots) and outside these plots.

Very little data is available on agriculture practised in residential plots. A team from the Projet Jardins et Elevages de Parcelles (JEEP) under the direction of Professor Jacques Paulus sj. has published articles summarising the data on this type of agriculture in Kinshasa. Such is the case of the article by J. Paulus et al (1989), which summarises the role of plot gardens and animal husbandry in the urban food supply in Kinshasa. Kabeya et al. (1992) presented data from the 1992 annual report of the NGO Jardins et Elevages de Parcelle. Kabeya et al (1994) presented data from the 1994 annual report of the same project. Mutuba et al, (1994) reproduce those of the 1994 annual report and the publication by Kabeya et al, (1994) gives the results of an inventory of the flora of housing plots. Monama et al (1985) published an article on the trophic chain of lead, taking into account the Matete mine production site. Makumbelo (1999) and Makumbelo et al. (2002, 2004) only looked at the situation in the Commune of Limete. Reading these and other publications can shed light on and provide in-depth information about urban agriculture in Kinshasa.

The National Extension Service of the Ministry of Agriculture and Community Development, supported by the FAO, has several brochures to encourage this type of farming. These include Guide de vulgarisation n°1-Culture vivrieres, n° 3 - Culture maraichere, n° 5- Petits elevages (Ministry of Agriculture and Community Development, SNV, 1992, 1993, 1994).

Data on agricultural activities carried out outside residential plots are shown in table 4b.

Table 4b. Data on agricultural activities carried out outside the housing plot.

Production site (Sourced ' information)	Date creation or start-up	Important dates and/or events (authorisation text)	Number of current farmers	Types of farming practised	Types of activities organised
Rice Kingabwa (farming) (1,2, 3)	1975- Chinese Agricultural Mission (MAC)	1982 by PNR is responsible for seed production. 1994 NRP involves international cooperation (None)	Around 5,000 farmers	- rice-growing farmer - seed production	Rice-mariculture (in season) dry)
Limete agricultural farm (1.6)	1975 by 58 rice-breeders- fish farmers	1994la farm loses around 4 ha to construction s anarchiques (None)	8 fish farmers rice farmers - fish farmers - stockbreeders	Rice - market gardening - livestock farming	Rice - market gardening - livestock farming
Market garden centre Echangeurde Limete (1,3)	1985-1986 visit functionair esdela Urban division of	(None)	50 market gardeners	- local market gardening	market gardening

	agriculture				
Technical Institute Agricultural Mombele (ITAM) (4)	1983 by Belgian Technical Assistanceand ITAM staff	1997un market gardeners group occupies part of the land	30 market gardeners	Market gardening	market gardening
		(None)			
Centre d'Accueil et de Passage de Limete (CAP) Social affairs (1)		Occupation by disabled people who grow vegetables (None)	7 market gardeners	Market gardening	market gardening
Lelongdu Boulevard Lumumba, Avenue Universiteet HGV road		(None)	35 market gardeners	Market gardening	market gardening

Lëgende : Sources of information (1) = Lokufa Batezwaka, Inspection du DR- Commune de Limete ; (2) = Tumba Langhe, rice farmer and President of the Cooperative agro-pastorale de Limete ; (3) = Kinika Nkuna, market gardener at the Centre maraicher de l'Echangeur de Limete ; (4) = Biese, President of the Comite et Intendant a l'ITA Mombele ; (5) = Masamba Ndombasi, VicePresident of CRK (Cooperative des Riziculteurs de Kingabwa); (6) = Lufila, President of the Limte Farm Committee; (7) = Theophile Lunhande, Secretary General of IDECOMI; (8) = Mukasie Luwezo Samuel, Manager of Soyapro ONG; (8) = Kanguele Mpili, Head of the CDI-Bwamanda incubation and pre-breeding centre; (10) = Dr. Vangu, Head of Department at CDI-Bwamanda. Vangu, Head of Service at CDI Bwamanda ; (11) = Madame LECOUTURIER, Soyapro Manager ; (12) = Marie Jose Mukeni, Soyapro ONG ; (13) = Dr. J.Paul Pierre, General Administrator.

2.2. The state of food security and urban agriculture

2.2.1. Protein-calorie malnutrition

2.2.1.1. Previous investigations

Sectoral surveys have been targeted more at the outskirts of the city. Also, since the looting of September 1991, the NGO Medecins Sans Frontiere (MSF) / Belgium and the Centre National de Planification de Nutrition humaine (CEPLANUT) have been regularly collecting data on the nutritional status of children of preschool age in the town.

In collaboration with the FAO and UNICEF, CEPLANUT is carrying out nutritional surveys in many sites in the capital to determine the prevalence of malnutrition.

The maternity wards of the Archdiocese of Kinshasa, managed by the Diocesan Office of Medical Works (BDOM), have been recording monthly data on children born with pelvic insufficiency for several years (DDOM, 1999). The classification criteria most commonly used by CEPLANUT and MSF for the height/weight ratio at -2 standard deviation (SD) refer to those adopted by WHO (1983, taken up by Toko and colleagues in CEPLANUT / FAO, 1994).

The height/age, weight/height and weight/age indicators are the most widely used.

2.2.1.2. Data on the city of Kinshasa

2.2.1.2.1. Chronic malnutrition

Surveys carried out in some districts of Kinshasa (CEPLANUT / FAO, 1994) gave in September 1992, on a population of 1,847, the prevalence of chronic malnutrition of 29.8% and in September 1993, on a population of 1,843 the prevalence of 29.8% (criterion - 2 ET). In 1996, the survey тепёе by CEPLANUT in collaboration with UNICEF (CEPLANUT / UNICEF, 1996) revealed pockets of acute malnutrition reaching 32.9% in eccentric towns and 42.9% in semi-rural towns.

2.2.1.2.2. Acute malnutrition

The CEPLANUT / FAO survey (cited above) found acute malnutrition rates of 5.2% in September 1992 on a population of 1,847 and 4.3% in September 1993 on a population of 1,843. The CEPLANUT/UNICEF (1996) survey (mentioned above) found pockets of acute malnutrition reaching 8.1% in semi-rural towns using the same criteria. UNICEF (2022) reports a percentage of 70% of malnourished children (www.unicef.org > drcongo > children).

2.2.1.2.4. Global malnutrition

Prevalences of 22.4% were found in September 1992 (CEPLANUT / FAO, 1994) and 22.8% in September 1993. In 1996 there were pockets of up to 34.1% of children in some semi-rural areas.

It should be remembered that the evolution of proto-energetic malnutrition varies over time, depending on its form. But the prevalence rates of global and acute malnutrition show an upward trend in both urban and rural areas. A study carried out in Kinshasa (1986) showed a global malnutrition rate of 22.0% and an acute malnutrition rate of 4.7%. These rates rose to 27.9% and 8.9% respectively a decade later (CEPLANUT / FAO, 1994). By 2022, an estimated 2.8 million people will be suffering from global acute malnutrition, including 1.2 million children under the age of five (HRP, 2022).

2.2.1.2.5. Low birth weight and brachial perimeter

No previous data produced on the basis of these two criteria is available.

2.2.1.3. Municipality data

No data on the nutritional status of children of preschool age (0-5 years) for the period prior to this survey were found for the Commune. The results of the few surveys that did include children from a few neighbourhoods were survived without much interest.

2.2.2. Food safety

2.2.2.2. Overview of the city's situation

The food security of the population of the city of Kinshasa can be assessed on the basis of data on the food security of households drawn from a defined sample. Any assessment of food security requires detailed data on food availability and the purchasing power of families at local level. Such data is not currently available (Ministry of Agriculture, Rural Animation and Community Development, 1991). For this reason, it is necessary to use other parameters, in particular the quantity of food consumed or available within the household (CEPLANUT, 1996). According to Jacqueline Andrd, quoted by Srinshaw et al (1991), the National Research Council's Food and Nutrition Bord's significant formula of 1 gram of protein per kg of body weight per day, of which 40% should be of animal origin and 60% of vegetable origin, is still the easiest for consumers to apply. Despite the standards established in many studies, there is unfortunately little data available to describe or define the problem in the city (Ministry of Agriculture, Rural Animation and Community Development, 1991). Nevertheless, the combined conclusions of the studies on   :

- daily calorie intake per person per day in households in certain areas of the   city (Bumbu and   Kingasani   );
- the frequency of family meals in the 12 neighbourhoods of Kinshasa, the stock of basic foodstuffs of at least 3   days   in the families of 12 neighbourhoods   ;
- The city's production of the main basic commodities confirms that the population of Kinshasa lives from day to day without being sure of tomorrow (CEPLANUT, 1996).

The people of Kinshasa get most of their calories and proteins from cassava in its various forms, rice, vegetable, fish, meat, fruit and other foods. Cassava alone accounts for more than half the weight of all the food consumed in the households studied so far (CEPLANUT, 1996). They make up more than 60% of the calories in the daily diet. Fresh vegetables provide a significant proportion of the population's protein intake.

2.2.2.3.   Food security in the Commune

There is no structure as such responsible for household food security in the Commune. Nevertheless, the partners involved in urban agriculture are also indirectly encouraging the availability of food in households.

2.2.3. Urban agriculture

2.2.3.2.   Overview of the city's situation

The desire to supply the town with fresh vegetables dates back to 1950. This prompted the colonial authorities to launch the N'djili Valley Development Project in 1954. In 1957, Kimbanseke began growing "European-style" vegetables. The political upheavals of 1960 meant that it was not until 1965 that, with the help of the French Cooperation, the public authorities decided to improve the Kinshasa market belt. In 1967 the marshland belt was extended. CECOMAF, a support structure for the marketing of market garden and fruit products, was set up in 1972. In 1979, the Market Garden and Fish Farming Project (P.M.P.) was set up. In 1981, an extension was built in the Mokore and Bono areas. It was financed jointly by the Public Treasury, France and the EEC. Work was completed in 1986.

The first attempts to grow rice began in 1974. They were extended in 1980 with a 1,000 ha programme by the Chinese Cooperation.

In 1972, 120 gardens were laid out around the Limete interchange, covering an area of 8 ha (usable agricultural area). The technical level of the market gardeners was quite high, but they needed to be supervised. The area was under constant threat of urbanisation (Département de l'Agriculture, 1987). As a result, vegetable and other production sites are being laid out across the town. Plots of housing, the city perimeter and along major roads are invaded by spontaneous agricultural activity.

In 1980, the volume of fresh vegetables produced in Kinshasa reached 20,000 tonnes (Department of Agriculture, 1987).

Fourteen years later, the city's production of the main staple foods is as shown in Table 5.

Table 5. Production of main staple foods in the city

Denreesde base	Hectares	Production in tonnes

Cassava	2.920	15.040
But	20.557	11.729
Rice	3.722	6.700
Beans	2.111	900
Sweet potato	685	2.810

Source: Department of Agriculture: Yearbook of Statistics reproduced by CEPLANUT, 1996. There is also a significant supply of vegetables from market gardens throughout the city, as well as from the gardens of residential plots (CEPLANUT, 1996). Urban farmers also produce fish, mushrooms, soya juice and wine. This spontaneity on the part of urban farmers continues to worry many a land developer, given the location of these sites, some of whose foodstuffs appear to be highly exposed to various forms of pollution (Monama et al., 1985).

2.2.3.3. Different types of urban agriculture practised in the Commune of Limete

Two types of agriculture can be observed throughout the Commune. These are : - professional urban agriculture practised on production sites outside the   plots and

- of plot-based agriculture carried out on residential plots (whether inhabited or not) and the surrounding area.

2.2.3.2.1. Professional urban agriculture

This type of urban agriculture is essentially characterised by :
- a large operating area ;
- a large production ;
- intensive, large-scale cultivation and/or livestock farming geared more to the imirelie ;
- a professional farmer who knows and manages the farm.

These farms are found :

Rice and vegetables are grown in the valley, around the town and along the main roads:
- Kingabwa rice-growing (rice-growing-mariculture) ;
- at the Limete agricultural farm;
- at the Echangeur market garden centre ;
- at the Institut Technique Agricole de Mombele ;
- along major roads, notably Boulevard Lumumba, Route de Poids Lourd and Avenue Universite-Sendwe;
- at the Centre d'Accueil et de Passage (CAP)-Affaires Sociales.

For fish farming and livestock rearing on the farm and other sites:
- practical fish farming and breeding on the Limete agricultural farm;
- livestock farming organised on avenue Kimbau by the Cooperative agro-pastorale de Limete ;
- poultry production at the Kimbanguiste Centre ;
- multiplication of poultry broodstock at the CDI- Bwamanda incubation and pre-breeding centre;
- Soya and mushroom-based by-products at SOYAPRO ;
- the production of poultry feed and eggs organised by the Societe Agricole et Veterinaire (SAVET).

2.2.3.3.2. Occasional farming

Occasional farming is the type of farming practised on small areas of housing plots (inhabited or not) and the surrounding area. Several publications provide valuable information. These include Paulus et al (1989) and Kabeya et al (1994). We will come back to them in the following lines.

2.2.3.3. Support structures for urban farmers in the Commune

Although urban agriculture in the commune did not attract the attention of any urban agriculture supervisors immediately after the country gained independence, the years 1975-1990 are recognised as those of urban agriculture supervision. These years were marked by the creation of several structures (essentially farmers' associations, development NGOs and other public structures). These partners are the same ones who oversee food security in the Commune.

Table 6 gives details of the Commune's support structures.

Table 6. Management structure for urban agriculture in the Commune.

Support partners	Natureof partner (year creation)	Framing radius	Type of farmers supervised
Limete agro-pastoral cooperative	Farmers' association (1974)	Kingabwa, Nzadi, Ndanu neighbourhoods	Rice growers
Limete agricultural farm	Farmers' Association (1975)	Salongo industrial district	Rice growers, breeders and fish farmers
Soyapro-ONG	ONG-D (1975)	Neighbourhoods Nzadi, Mbamu	Parishioners, Women united inAMEK , Farmers whose children are malnourished
Centrede integrated development-CDI Bwamanda	ONG-D (1976)	The whole city	Poultry farmers
Gardens and Breeding Plots - JEEP	Project (1988)	Neighbourhood Mbamuet Nzadi	Households with a malnourished stage III child, neighbours and others residentsof
			district
IRES Theatre	ONG-D (1989)	The whole city	Farmersof media-urbanand populations disadvantaged by the theatre
Presbyterian	Programme	Mbamu district,	Households with a

gardening programme- Presbyterian Community of Kinshasa - PPJ/CPK	(Missionary Service) (1989)	Mfumu N'Vula	child malnourride stageIIwhich visit C.S. Mayamba and Libiki
Initiation to integrated development- IDECOM-ONG	ONG-D (1990)	Kingabwa yaounde, Kingabwa village Mandrandela and Pecheurs neighbourhoods	The farms farmers, market gardeners, livestock breeders plots, the fishermenand other
Support programme for mothers bongisa and the follow-up of malnourished children in stage II BDOM-NGO	Programme (Missionary Service) (1993)	Mombele, Mfumu Nvula, Kingabwa yaoundeet neighbourhoods Mbamu	Households with a child malnourride stageIIwhich frequent the parishes ofSt Felix, St Amand, Gonza and Kizito
Service l'agriculture, Servicedu Developpement Rural, Service de l'Environnement	Public partners	All of Municipality	Agriculture

These structures are supported by incentives, techniques and associations. The majority of farmers work individually, using conventional techniques with little rationality. This partly explains the meagre harvests, which have not been quantified.

Chapter III. METHODOLOGICAL APPROACH

3.1. Approach

Le travail a ийНзё 1 approche systëmique (ou synthetique) qui consiste a considërer un systëme complexe par ses carac^ristiques ëmergentes Нёез a sa totals, propriëtë qui n'est pas reductible a une addition des caracteristiques de ses elements (Sciwarz, 1997 cite par Gobal et al., 2003). It is the science of describing and explaining the diversity^ of forms (Lacoste and Salanon, 1999). It is also an approach that emphasises the links and interactions between parts (Giordio and Saltet, 2011 cited by Makumbelo et al., 2023). The systëmique approach nommёe aussi analyse systëmique is an interdisciplinary field relating to the l^tude of objects in their complex^. (en.m.wikipëdia.org).

The urban agricultural production system is subdivided into 6 sub-systems (Maldague, 1996 taken up by Makumbelo, 1999). Figure 2 shows the overall approach to this system.

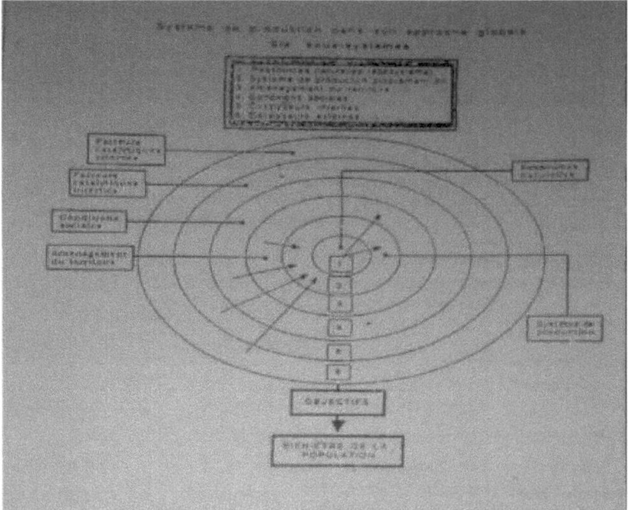

Figure 2. Production system in its global approach (Maldague, 1996 taken up by Makumbelo, 1999).

From this figure, it can be seen that the system under study comprises :
-    natural resources (different agricultural production sites) ;
-    the production system itself (the land with its land tenure problems and its availability: the production forces with the different types of social production system; the production capacity of these different players and the inputs, essentially the goods and production factors available);
-    regional planning (the necessary infrastructure and facilities, which in this case are essentially the subdivision and location of these production sites in relation to the causes of pollution from production).
-    the social conditions of producers (the health of farmers defined by the nutritional level of their children, the level of availability of food to mënages, Гaccёз to support, training and technical supervision ;
-    the catalysts, which are the Commune's agricultural and environmental support services, the Coopëratives and other farmers' associations (internal catalysts). The NGDOs, the action of international bodies (UNICEF, European Union, ADF, WFP, Embassies (external catalysts). The combination of all these sub-systems and ëlëments of the system ёme emerge in an objective which is the Well-being of the population.
This work, which has a clear objective, retains only what is essential to respond to its preoccupation.

3.2. Equipment
*** For all surveys**
-    a geographical map ;
-    plot numbers to be surveyed
*** For nutritional aspects**
-    a salter scale ;

-    a "Rolamëtre" measuring rod;

-    a wooden measuring rod and a non-blastic strip with 3 marks and coloured successively in red, yellow and green;

-    the tables from "Measuring change in nutritional statuts-guide line for vulnbrable group world a l'bcart type (-2ET).

*** For food security and urban agriculture**

-    a questionnaire;

-    a logbook ;

-    a pen and tablet.

3.3 Data collection

*** For nutritional aspects**

-    Weigh children under 5;

-    measure the height of the child lying on the Rolambtre bed or standing up using a wooden height gauge;

-    lift the upper arm circumference ;

-    ask mothers about the opportunities they hope to improve their children's nutritional status.

*** For food security and urban agriculture**

-    measurement of gardens encountered ;

-    identification of the types of resources (plant or animal) grown (if in doubt, take samples for better identification) ;

-    estimate the quantity of production.

3.4 Interpretation of results

*** Reference indices for nutritional aspects**

-    weight sampled in grams ;

-    marked size in cm ;

-    length of brachial perimeter recorded in cm :

-    between 12.5 cm and 13.5 cm (yellow part) = slightly malnourished ;

-    less than 12.5 cm (red part) = severely malnourished;

-    over 13.5 cm (green part) = well nourished - underweight (maternity statistics).

*** Benchmark indicators for the state of food security and urban agriculture**

- farmers' expectations to ensure food security for their households; - availability, accessibility of permanent fagon.

All the data will be grouped together in tables to show the trends (characteristics) of each stratum of the sample.

3.5. Identification of species

Botanical identification of species is the starting point and one of the pillars of the basis of work (Mitja, 1992) in ethnobotany. To do this, it was first necessary to collect samples of all the species mentioned in the survey. The results were then passed on to a sample of the population, so that specimens of the species mentioned could be collected for identification. This was made possible by consulting various floras (Liben, 1987; Pauwels, 1983, 1992; Lebrun and Stork, 1991, 1992,   1995, 1997). This documentation has been enhanced by the indicator service

botanicals from the INERA herbarium.

3.6. Calculating production

For vegetable production, the quantity was estimated by measuring the surface area containing the vegetable and weighing a sample of the produce.

For fruit production, the quantities were estimated by a sample of owners who noted the quantity collected each time it was picked. The average quantities reported by the households were compared with those reported in the literature for each species. A number of publications (Van Den Abeele and Vandenput, 1956; Vandenput, 1981; Tazenas du Montcel, 1985; Anonymous, 1989) were used as support.

The assessment of the annual quantity of fruit produced and the availability of fruit per species, per person and per day was based on previously published formulae (Makumbelo et al., 2005), which are no longer reproduced in this book.

3.7. Survey of a sample of residential plots

3.7.1. Sampling

The concept sample derives from the statistical principle that a part of the universe exhibits the basic characteristics of the universe, when the selection of the ëlëments of the sample is made in an appropriate and rigorous iaeon. A sample is a part of the population constituted by a chosen process, in дёпёгаl, of dëlibërëe iaeon for the purpose of studying the properties of the original population (Ekofo, 1989; Wong et al., 2001; Makumbelo, 2023; Makumbelo et al., 2023).

The sample study has two essential aspects: the mathematical aspect (which enables the sample to be constituted and the possible error to be calculated) and the psychological aspect (consisting of the drafting of the questionnaire, the choice of interviewer training, the type of survey monitoring and the identification of the espëces ressources dënombrëes) (Essanga, 1989; Makumbelo, 2023; Makumbelo et al., 2023).

3.7.1.1. Mathematical aspects

3.7.1.1.1. Sampling

Sampling is an essential technique used for economic reasons. Since the population to be surveyed is generally too large to be completely surveyed, only a portion of it is used (FAO, 1981). It is the representative subset of the ëtudiëe population that has consented to the main characteristics of the latter (Muluma, 2003).

The sample covers residential plots in the districts of the Commune of Limete.

3.7.1.1.1.1. Statistical unit or survey unit

This study uses an objective sample survey (Wong et al., 2001). It uses the 'plot' as the 'sampling unit' and the 'household' as the 'statistical unit'. The plot is defined as an allotted, bounded piece of land containing at least one dwelling house and one or more households. The household comprises a group of people who share their main meal together, live together under the same roof and are under the authority of a single head of household.

Consultation of administrative census forms at Commune offices and neighbourhood offices, and systematic counting of plots of land, has made it possible to compile an accurate plot register (Makumbelo et al., 2002, 2007) such as that used by the Institut National de la Statistique (INS, 1989).

The surveys were carried out on a random sample of all the plots according to Wonnacott et al. (1991) and stratified (Norma et al., 1978; White et al., 2001; Wong et al., 2001; Makumbelo et al., 2002, 2007).

3.7.1.1.1.2. Stratification

Stratification is the process which ensures that all the characteristics of the population can appear in the sample. It is the population's index of reliability.

The stratification is defined with reference to Makumbelo et al (2002) and Makumbelo

(2023), using the age and standing of the entities or neighbourhoods as criteria. These criteria are presented in table 7, for age (character of what has existed for a long time) and for standing (comfort, standard of living and social rank).

Table 7. Strata and stratification criteria.

Strata	Stratification criteria	Criteria
I	Former high-standard property	A.H.
II	Former mid-range property	A.M.
III	Former low-rise property	A.B.
IV	Moderately old property with high standing	M.H.
V	Medium-anciententitea Medium standing	M.M.
VI	Moderately old low-standing property	M.B.
VII	Recent high-standard property	R.H.
VIII	Recent mid-range property	R.M.
IX	Recent low-rise property	R.B.

The age of the districts is defined in relation to the time when the entity was recognised as an administrative district. Thus we have old districts (I) established as administrative districts before independence until 1973. Moderately old districts (II) resulting from the enlargement of administrative divisions in 1974, 1975 and 1976. Recent districts (III) made up of entities created after these divisions.

Maman Nzenze is considered a recent administrative district simply because it deals administratively with the Commune without an intermediary.

The neighbourhood's standing is determined by the following indicators:

1° All households are supplied with water through the REGIDESO connection.

2° Each plot has its own system for connection to the water distribution network.

3° Each plot is inhabited on average by one household which has its own sanitary facilities incorporated into the dwelling house.

4° The majority of households have their own means of transport.

5° The houses on the majority of plots are built from durable materials and have enough space for the members of the household.

6° These plots are fenced with durable materials following a certain architectural style.

7° Most of the vehicular access routes in the neighbourhood are maintained.

8° the entity has public lighting maintained by SNEL.

The combination of these criteria (age and standing) and the corresponding stratification indicators was used to define the following 9 strata (Table 8).

Table 8. Different strata, stratification criteria and indicators.

Strata	Stratification criteria and indicators
Stratum I Stratum II Stratum III Stratum	High-standard old neighbourhoods (A.H.). These are group I areas where the vast majority of plots, houses and households meet indicators 1, 2, 3, 4, 5, 6, 7 and 8. Old middle-class neighbourhoods (A.M.). These are the entities in group I and about half of the plots, houses and households meet the indicators: 1, 2, 3, 4, 5, 6, 7 and 8.

IV	Low standard old neighbourhoods (A.B). These are group I properties, very few of which meet the indicators: 1, 2, 3, 4, 5, 6, 7 and 8.
Stratum V	Quartiers Moynement Anciens a Haut Standing (M.H.). These are group II areas where the majority of plots, houses and households meet the indicators: 1, 2, 3, 4, 5, 6, 7 and 8.
Stratum VI	Middle-aged neighbourhoods of medium standing (M.M.). These are group II areas, with around half of the plots, houses and households meeting the following indicators: 1, 2, 3, 4, 5, 6, 7 and 8.
Stratum VII	Middle-aged, low-status neighbourhoods (M.B.). These are group II areas where very few of the plots, houses and households meet the indicators: 1, 2, 3, 4, 5, 6, 7 and 8.
Stratum VIII	Recent high-standard neighbourhoods (R.H.). These are group III units, with the majority of plots, houses and households meeting the following indicators: 1, 2, 3, 4, 5, 6, 7 and 8.
Stratum IX	Recent medium-status neighbourhoods (R.M.). These are group III entities, with around half of the plots, houses and households meeting the indicators: 1, 2, 3, 4, 5, 6, 7 and 8. Recent low-status neighbourhoods (R.B.). These are group III entities, with very few plots, houses and households meeting the indicators: 1, 2, 3, 4, 5, 6, 7 and 8.

This table shows that the combination of two criteria and the corresponding indicators produced 9 different strata.

3.7.1.1.1.3. Grouping entities into strata

Most of the entities show a degree of internal homogeneity in terms of these criteria and indicators. This makes them easy to stratify, despite some minor differences.

After arranging the various entities into strata, the grouping of entities into strata is as follows (table 9).

Table 9. Grouping of entities by stratum.

Strata	Criteria	Entities
I	A. H.	Limete residential, Limete industrial, (old part), Kingabwa industiel.
II	A. M.	Kingabwa Yaounde, Kingabwa village.
III	A. B.	Mombele.
IV	M.H.	Salongo (industrial).
V	MR M.	Mbamu, Nzadi, Salongo (non-industrial), Mayulu.
VI	MR B.	Mososo, Mateba, Mfumu Nvula, Agricultural, General Masiala..
VII	R. H.	Kingabwa Mandrandele.
VIII	R. M.	Not represented and therefore not included in the tables.
IX	R. B.	Ndanu, Maman Nzenze, Limete industriel (new part-Bobozo and annex), Kingabwa pecheur.

Legend: Criteria = A: Old, B: Low, H: High, M: Medium, R: Recent.

This table shows that none of the neighbourhoods surveyed meet the criteria for stratum VIII: Recent to Medium Standing.

3.7.1.1.1.4. Drawing the sample

Based on a plot-by-plot survey of 18,480 plots, a stratified sample was drawn as shown in Table 10.

Table 10. Sample drawn.

Strata	Total plots surveyed	Proportion of plots selected from the sample	Sample
I	2.274	28.4	28,0
II	1.000	12,5	13,0
III	883	11,0	11,0
IV	160	2,0	2,0
V	4.267	53 ;3	53,0
VI	5.529	69 ;1	69,0
VII	583	7,3	7,0
VIII	+	+	+
IX	3.779	47,2	47,0
Tot	18.480	231,0	231,0

Lëgende = + : nothing to report.

Table 10 shows that the sample drawn, with a sampling percentage of 1.25%, i.e. a fraction of 1/10, consists of 231 plots to be surveyed.

3.7.1.1.1.5. Validation of results

The part drawn in the sample P = 1.25% and not drawn Q = 98.75% there is a 99.7% probability (area corresponding to + 3 times the deviation) that the mean of the universe falls between the following limits: a = Jl'25-98.75-(referring to CEPLANUT, 1988 taken up by Makumbelo, 1999).

V 18.400

= 0,08

3o = 0.08 x 3

= 0,24

a. 1.25 - 0.24 = 1.01% lower limit

b. 1.25 + 0.24 = 1.49% upper limit with a margin of error of ± 0.24%.

3.7.1.1.1.6. Tabulation and presentation of results

As mentioned above, the data must be compiled in tables to show the disparities between the different strata.

In the Results and interpretation section of this work, all the results from the stratified sample are grouped by stratum (I to IX) in the various tables. The only exceptions are tables 19, 20, 27, 34 and 37.

The count was carried out manually, using a calculator to perform the various operations.

3.7.1.2. Psychological aspect

3.7.1.2.1. Designing and drafting the questionnaire

The questionnaire consists of sending a large number of people the same list of questions prepared in advance, followed by the systematic dëpouⅢement of the answers having ëlëments for which the opinion, judgement or Involution of a subject interrogatedë are sought. It is written using the imagination and intuition of the researcher.

The questionnaire *was* designed to find out, through knowledge of the population in their environment, their expectations regarding health care for the members of their households, the activities organised by the household to achieve this, the inputs that the members use and the outputs that they obtain. This questionnaire did not forget where these activities take place.

3.7.1.2.2. Administration of the questionnaire

Being in principle anonymous and self-administered, the investigator only intervenes in cases of extreme necessity. Here he is prepared to be sent to defined groups, considered to be inhabitants of Kinsiasa, who either engage in urban farming or have children under the age of 5 to weigh. Urban farmers are either those who farm in their residential plots, or those who farm at the Kinsiasa provincial office.

The mode of questionnaire self-administration was determined, with reference to Makumbelo et al, (2002, 2007, 2008) and Makumbelo et al, (2016), by the results of the pretest. These results dictated a particular choice of interview and type of questionnaire self-administration. Indeed, although it is recognised that the interview, in the technical sense, can be defined as a scientific investigation procedure, using a verbal communication process, to gather information in relation to the set goal (Muluma, 2001), the population of Kinshasa being very preoccupied with finding solutions for survival, verbal communication is more appropriate than sending a questionnaire to be completed alone by the respondents.

If questionnaire self-administration consists of the respondent reading (alone) the questionnaire before answering. In this work, the question was asked verbally to the respondent and the respondent's answers were recorded on the survey form without any further exchange between interviewer and respondent.

A plot includes all the people met at the time of the survey around an adult from the plot visited. The survey counts this plot as the survey unit and the adult (older) person met is taken as the person surveyed. The survey used an open-ended questionnaire.

It should also be pointed out that "the staple food is defined as the prepared food, which in a normal food ration, quantitatively exceeds all the other foods that are an integrated part of a ration (Department of Agriculture and Food).

Dëveloppement rural, 1987). In this work, the availability of staple foods is only indicative. These products are themselves purchased more than they are produced by households. Production techniques and their impact on the quality of each of these foods, as well as the conditions under which they are produced, have not been the subject of this study.

3.8. Other surveys that produced non-statistically tested data

This section includes interviews conducted without a predefined statistical sample. The quality of the data collected was confirmed by simple triangulation.

The surveys were carried out among modern producers, who are organised producer structures, and among professional farmers on sites other than housing plots.

3.9. Observation

From 1999 to the present day, 24 years on, continuous observation has been carried out. All new developments have been recorded.

II PART

PRESENTATION AND INTERPRETATION OF RESULTS

This part includes : Presentation of the results (chapter 4) and Interpretation of the data (chapter 5).

Chapter IV. PRESENTATION OF RESULTS

4.1. Data from households surveyed

4.1.1. Socio-demographic data

Data on the plots and households surveyed, the average number of households living in a plot and the population encountered by stratum are shown in Table 11.

Table 11. Location of plots surveyed.

Strata	Sample plots	Plots surveyed	Household surveys	Average number of households per plot	Population met
I	28	17	31	1,8	158
II	13	12	21	1,7	169
III	11	8	17	2,1	136
IV	2	2	4	2,0	32
V	53	53	122	2,3	1.006
VI	69	63	147	2,2	1.177
VII	7	7	10	1,1	80
IX	47	39	61	1,5	393
Total	231	201	413	2,0	3.151

This table shows that a sample of 231 plots was drawn from the survey of 18,480 plots. 201 plots were surveyed in which 413 households were counted, for an average of 2.0 households per plot.

Table 12 shows the number of children aged 0-5, the average (population per plot, per household and children aged 0-5), the number of children and the number of farmers per plot.

Table 12. Number of children under 5, average (of population per plot, per ténaдe and children under 5), number of children and plot farmers.

Number of children aged 0-5	Average population per plot	Average population per ménage	Average number of children aged 0-5 per household	Number of children aged 1-5 to be surveyed	Number of urban parcel farmers
18	9,2	5,0	0,5	9,0	19,0
18	14,0	8,0	0,8	9,0	11,0
20	17,0	8,0	1,1	10,0	8,0
6	16,0	8,0	1,5	3,0	2,0
98	18,0	8,2	0,8	49	38
109	18,6	8,0	0,7	55	125
15	11,4	8,0	1,5	8,0	9,0
70	10,0	6,4	1,1	35,0	41,0

354	15,6	7,6	0,8	178	253

From Table 12, it turns out that 354 children aged 0-5 years have ëtë dënombrës in a population average of 15.6 per plot and 7.6 per mënage. The average number of children aged 0-5 per mënage is 0.8. The number of children aged 1-5 years to be surveyed is 178 and the number of plot farmers is 253.

4.1.2. Characteristics relevant to the nutritional survey

Table 13 shows the characteristics of the nutritional survey.

Table 13. Situation of children aged 0-5 by stratum.

Parcels investigëtëes	Number of children aged 0-5 years recensës	Number of children aged1-5 ansa investigate	Number of children aged 1-5 effectively enquëtës
17	18	9	9
12	18	9	6
8	20	10	10
2	6	3	0
53	98	49	41
63	109	55	44
7	15	8	7
39	70	35	31
Total	353	178	148

Out of a total of 353 children aged 0-5 years, 178 children aged 1-5 years were identified and 148 were actually weighed and their height measured.

4.2. Data on nutritional status

4.2.1. What mothers hope to achieve to improve their children's health

The mothers' responses with a view to improving their children's health are recorded in table 14.

Table 14. What mothers hope to improve their children's health.

Number of mothers surveyed	Good nutrition	Infection control	Controlling social problems	Coping with family problems	Abstention
9	7	1	1	0	0
6	4	1	1	0	0
10	4	2	2	2	0
0	0	0	0	0	0
41	14	6	9	3	9
44	21	13	13	0	0
7	3	2	1	1	0
31	10	7	7	4	3
Total	63	32	34	10	12
%	42,5	21,6	22,9	6,7	8,1

Out of a total of 148 mothers (of children who weighed), 42.5% thought that with good nutrition they could ensure their children's growth. 22.9% believe that mastery of the social (financial, political, etc.) environment will enable them to control the conditions for their children's healthy growth.

4.2.2. Data on the nutritional situation (malnutrition in children aged 1-5).

The situation regarding chronic malnutrition (T/A) is shown in table 15.

Table 15. Number of children suffering from chronic malnutrition.

Number of children aged 15 measured	Malnourished children	Well-nourished children
9	5	4
6	2	4
10	2	8
0	0	0
41	13	28
44	12	32
7	2	5
31	13	18
Total	49	99
%	33,1	66,8

The results recorded in this table (15) show that 33.1% of children considered to be chronically malnourished.

Table 16 shows the percentage of children suffering from malnutrition; iiguc' (P/T)

Table 16. Percentage of children suffering from acute malnutrition.

Number of children aged 1-5 years weighed and height measured	Malnourished children	Well-nourished children
9	6	3
6	0	6
10	0	10
0	0	0
41	1	40
44	3	41
7	0	7
31	7	24
Total	17	131
%	11,4	88,5

Table 16 shows that 11.4% of children are malnourished. No cases were recorded in strata II, III, IV and VII.

Table 17 shows the percentages of children suffering from malnutrition as measured by brachial përimëtre.

Table 17. Children suffering from acute malnutrition (brachial perimeter).

Children whose brachial perimeter measures	Severely malnourished children (nR)	Moderately malnourished children (nJ)	Well-nourished children (nV)
9	3	3	3
6	0	1	5
10	2	1	1
0	0	0	0
41	5	1	29
44	1	8	35
7	0	0	7
31	8	6	17
Total	19	26	103
%	12,2	17,5	69,5

Analysis of table 17 shows that 30% of children suffer from mgim malnutrition, compared with 69.5% of well-nourished children.

Table 18 shows the percentages of children suffering from global malnutrition.

Table 18. Percentage of children suffering from global malnutrition (G/M).

Number of children aged 1-5 weighed	Children suffering from malnutrition	Well-nourished children
9	5	4
6	3	3
10	3	7
41	14	27
44	13	31
7	0	7
31	16	15
Total	54	94
%	36,0	63,5

Table 18 shows that 36.0% of children are malnourished compared with 63.5% in good health.

Table 19 shows the percentages of children born with pelvic insufficiency (at the Kingabwa 1 and 2 and Musoso maternity hospitals in the 1[er] quarter of 1996).

Table 19. Situation of children who were born with underweight.

Month	Total deliveries in Kingabwa	Birth with < 2500g in Kingabwa	Total Musoso births	Birth with < a 2500ga Musoso
January	225	37	41	2
February	200	40	54	5
March	260	36	41	0
Total1[er]	685	113	136	7

quarter				
%	100	16,0	100	5,0

Source: Quarterly report from Liziba Hospital. Z.S. Kingabwa-Archidiocese of Kinshasa (BDOM).

The table above shows 16% and 5% of children born with pelvic insufficiency at the Kingabwa and Mososo maternity hospitals respectively.

4.3. The state of food safety

4.3.1. Parents' expectations

The expectations of parents to ensure the security of their households are expressed in table 20.

Table 20. Main ways in which parents hope to ensure food security for their households.

Agricultural production	Financial resources	Agricultural production and financial resources	Hope in God	Debrouilla rdise	Inconsistent responses
122	81	5	13	17	15
46,2%	32,0%	1,9%	5,1%	6,1%	5,9%

From the analysis of Table 20, it emerges that 46.2% of mënages 1^përeΠ by agricultural production, 32.0% by financial means and 1.9% by agricultural production and financial means.

4.3.2. Availability of plant resources and livestock products produced by households.

The information relating to this point is shown in table 20.

Table 20. Availability of livestock resources and livestock products produced by households.

Mënages farmers	Disponibil rCë up to 3 days with a vegetable garden	Disponib ПИё up to 3 days with fruit tree	Disponib ПИё up to 3 days with ëlevage	Disponibil Иё of more than 3 days with vegetable garden	Disponibil Иё of more than 3 days with fruit tree	Disponibil Иё of more than 3 days with ëlevage
16	7	11	4	2	4	2
11	5	7	3	5	4	0
9	3	3	3	2	5	0
2	1	2	0	1	0	0
48	24	36	7	15	8	7
55	19	50	17	14	2	2
7	3	6	1	4	1	1
36	15	11	13	16	22	5
Total	77	126	46	60	46	17
%	18,6	30,5	11,6	14,5	11,1	4,1

Analysis of Table 20 shows that the availability of food is more assured by fruit trees and vegetables than by livestock. This is the case both for up to 3 days and for more than 3 days a

37

week.

4.3.3. Availability of staple foods (starchy foods) most consumed by households

Table 21 shows the availability of staple foods. This table is cc^ë for reasons of space in Table 21a and 21b.

Table 21a. Availability of staple foods per mënage.

Cassava		Cassava and maize		Cassava, chikwangue and malemba		Cassava and rice	
(1)	(2)	(1)	(2)	(1)	(2)	(1)	(2)
0	5	9	2	0	2	5	0
2	1	2	6	0	0	4	1
2	0	4	5	0	0	5	0
0	0	1	0	0	0	0	0
8	5	42	29	7	3	7	6
18	14	54	22	12	7	5	0
0	0	3	3	0	1	1	2
6	3	24	9	5	9	4	0
Tot. 36	29	139	76	24	13	31	9
%6 ,7	7,0	33,6	18,4	5,8	3,1	7,3	2,1

Lëgende : Tot. = total ; (1) = availability up to 3 days per week ; (2) = availability more than 3 days per week.

Table 21b. Availability of staple foods per mënage (continued).

Rice		Cassava, rice and maize		Cassava, rice and chikwangue		Maniocet plantain banana		Fruits of l' arbre a bread	
(1)	(2)	(1)	(2)	(1)	(2)	(1)	(2)	(1)	(2)
0	3	0	0	0	5	0	0	0	0
0	0	0	0	0	0	2	2	1	0
0	0	1	0	0	0	0	0	0	0
3	0	0	0	0	0	0	0	0	0
0	0	5	1	4	3	0	0	1	0
0	0	7	3	3	0	0	0	2	0
0	0	0	0	0	0	0	0	0	0
2	3	0	1	1	2	0	0	1	0
Tot. 5	6	13	5	8	10	2	12	5	0
% 1,2	1,4	3,1	1,2	1,9	2,4	0,4	0 ,4	1,2	0,0

Lëgende : Tot. = total ; (1) = availability up to 3 days per week ; (2) = availability more than 3 days per week.

An analysis of table 21b shows that out of 413 households surveyed, i.e. 100%, 33.6% have cassava available up to 3 days a week in the form of cossettes and mai's, 18.4% have it available for more than 3 days a week, 8.7% have cassava available up to 3 days a week and 1.2% have breadfruit *(Artocarpus altilis)* available up to 3 days a week.

Chikwangue and malemba are transformations of cassava.

4.3.4. Availability of food to households who farm in the valley, on the outskirts of the valley and along major roads

The food security of these households is taken into account in the food security of all households surveyed. It should be pointed out that around 5,001 farmers (100%), 8 fish farmers (100%) and 122 market gardeners (100%) had rice available for more than 3 days a week during the harvest period, as well as fish and various vegetables.

The production of CDI Bwamanda, SOYAPRO ONG and Sociëtë Agricole et Vëtërinaire (SAVET) is a supplement to the food available on the market.

4.4. Urban agriculture

The section on urban agriculture deals with the conduct and production of these activities (agricultural production on housing plots and agricultural production outside housing plots).

4.4.1 Farming activities

4.4.1.1. Agricultural activities carried out on residential plots (inhabited or not) and their surroundings.

Table 22 shows the distribution of farmers by age, gender and financial resources.

Table 22. Distribution of plot farmers by age, sex and financial means.

Total men	Total women	Menagesqui can be found at serious financial difficulties	Menagesqui have some financial possibilities	Menagesquiont plus financial opportunities
4	15	11	3	5
3	8	3	3	5
3	5	7	1	0
1	1	2	0	0
18	20	19	10	9
17	108	74	40	11
5	4	4	2	3
14	27	12	18	11
Tot . 45	188	132	77	44
%25 ,6	74,3	52,1	30,4	17,3

Legend: Tot. = Total.

Analysis of Table 22 shows that women are more involved in agricultural activities than men. More than half of households are experiencing serious financial difficulties.

4.4.1.2. Different agricultural activities organised by households

The different agricultural activities organised by households are grouped together in table 23.

Table 23. Agricultural activities organised by households.

Vegetable garden	Fruit trees	Breeding	Garden and fruit trees	Garden and livestock	Fruit trees and livestock	Garden, fruit trees and livestock	Ont of agricultural activities	Have no agricultural activity
1	6	0	3	0	1	5	16	15
0	0	0	8	0	0	3	11	10
1	2	0	3	0	0	1	9	8
0	0	0	2	0	0	0	2	2
2	3	1	29	1	4	8	48	74

1	15	1	20	1	5	11	55	92
0	0	0	5	0	0	2	7	3
0	2	2	16	1	1	14	36	25
Tot. 5	28	4	86	3	14	44	184	229
%2 ,7	15,2	2,1	46,7	1,6	7,6	23,9	100 44,5	55,5

Legend: Tot. = Total.

Analysis of table 23 shows that 44.5% of households surveyed organise at least one agricultural activity. Most households have a vegetable garden and at least one fruit tree. Very few households have a garden and livestock. The fruit tree is the most organised activity.

4.4.1.3. Location of vegetable gardens

Table 24 shows where the vegetable gardens are located.

Table 24. Location of vegetable gardens.

Number of gardens	Residential plots	Rue	Values	Other land	Plots and streets	Plots and values
9	8	0	(1)	0	0	0
11	10	0	0	1	0	0
5	4	0	(1)	0	0	0
2	2	0	0	0	0	0
40	32	0	(2)	0	0	3
33	27	6	0	0	0	1
7	5	0	0	2	0	2
31	28	2	0	1	0	0
Tot. 138	116	8	(4)	4	0	6
% 100	84,0	5,7	(2,8)	2,8	0,0	4,3

Lëgende : Tot. = Total ; ( ) = not seen during inventory.

Analysis of Table 24 shows that more than 80% of kitchen gardens are placës in residential plots and more than 5% in the street. Only 2.8% are placës in other plots which are ëссĸз, ë churches or other places close to the housing plot. No mënage has vegetable gardens placedës both in the housing plot and the street.

4.4.1.4. Features of kitchen gardens on residential plots and the surrounding area

The main characteristics of the gardens surveyed are shown in table 25.

For reasons of space, this table is dividedë into 25a and 25b.

Table 25a. Characteristics of the gardens surveyed.

Area of land occupied (in m $)^2$	Area of unoccupied land (in m $)^2$	Garden in the plot	Garden outside the plot	Garden fence	Unenclosed garden	Use of fertilisers (organic and chemical)	No use of fertilisers
213	360	8	1	2	7	6	3
276	318	10	1	0	11	5	6
77	250	4	1	1	4	3	2
100	44	2	0	0	2	1	1

1227	3594	31	8	2	37	22	17
931	2392	27	6	1	32	20	13
261	590	5	2	0	7	5	2
1074	4830	28	3	10	21	25	6
Tot. 4.159	12.476	115	22	16	121	87	50
%45 ,3	54,6						
%		83,5	16,0	11,6	88,3	63,5	36,2

Legend: Tot. = Total.

Analysis of part a of table 25 rёyёк shows that over 54% of plots have space to install a garden. These plots have 12,476 m^2 . Overall, 88% of plots are unfenced and 83.5% of gardens are laid out on residential plots.

Part b of table 25 contains the rest of the plot garden characteristics.

Table 25b. Characteristics of the gardens surveyed.

Gardens exposed to pollution	Jardins proteges dela pollution	Gardens watered by l'eau de la Regie des eaux	Gardens watered with a other water
1	8	6	3
3	8	8	3
1	4	4	1
1	1	2	0
6	33	26	8
7	26	26	7
2	5	5	2
15	16	28	3
Tot. 36	101	105	32
%26 ,2	73,7	76,6	23,3

Legend: Tot. = Total.

The second part of table 25 shows that over 73% of the plots are protected from pollution and over 76% are irrigated with Regideso water.

4.4.1.5. Features relating to the size of kitchen gardens on residential plots

The characteristics relating to the surface area of the gardens are shown in table 26. For reasons of space, this table will be divided into a and b.

Table 26a. Characteristics relating to the size of the gardens.

Surface area of gardens (in m)^2	Surface area of undeveloped gardens (in m)^2	Average surface area of gardens (in m)^2	Number of gardens whose area is between 15 and 39 m^2 (in m)^2	Surface area of gardens with area is between 15 and 39 m^2 (in m)^2	Uncultivated area of gardens where the area is between 15 and 39 m^2	Average surface area of gardens with area is between 15 and 39 m^2 (in m)^2

					(in m)^2	
3	50	3	7	170	110	24
30	126	6	2	46	80	23
6	90	2	1	21	40	21
0	0	0	1	30	42	30
74	1186	5	12	266	719	22
67	500	6	9	184	399	20
23	160	11,5	1	20	120	20
84	1110	7	6	135	350	22
Tot. 287	3222	40,5	39	872	1860	182
%			28,4			
X in m^2		5,7				22,7

Lëgende : Tot. = Total.

Analysis of table 26a shows that the average size of vegetable gardens under 15 m^2 is 5.7 m^2 and that of gardens between 15 and 39 m^2 is 22.7 m^2 .

Figure 3 shows a vegetable bed laid out in the valley.

Figure 3. Vegetable bed in the valley.

Table 26b shows the characteristics of vegetable gardens measuring more than 39 m^2 .

Table 26b. Characteristics relating to the size of vegetable gardens (continued).

Number of gardens whose surface area is greater than 39 m^2 (in m)^2	Surface area of gardens with dimension a 39 m^2 (in m)^2	Area of gardens larger than 39 m^2 no operated (in 2$_{m)}$	Average area of gardens measuring 39 m^2 (in m)^2	Average area per stratum
1	40	40	40	24
4	200	62	50	25
1	50	60	50	15
1	70	0	70	50

14	887	1197	63,3	31
12	680	858	56,6	28
4	218	410	54,5	37
14	855	2850	51	35
Tot. 51	3000	5477	445,4	206
% 37,3				
X in m^2			55,6	31

Lëgende : Tot. = Total.

Table 26b shows that the average size of gardens larger than 39 m^2 is 55.6 m^2 .

4.4.1.6. Vegetables not yet processed

Table 27 shows the indigenous vegetables found in home gardens.

Table 27. Lëgumes not yet valorisës.

Strata where the species is listed	Number of western plots identified the espëce	Local name	NomfrangaisName scientist
3	1	Kataba	Justiciastratasubsp insularia (T.Anders) J.K.Morton
3	1	Milembwa	*Talinum fruticosum* (L.) Juss. Syn. *Talinum triangulare* (Jacq) Willd
3	1	Mutamwa	*Grassocephalum sarcobasis* (DC.boj)) S Moore
9	1	Tidi	*Phytolacca dodecandra* L. Herit.

In addition to Kikalakasa (*Psophocarpus scandens* (Endl.) Verdc., foundë in plots with more than one stratum, a few other mdigënes tegumes have ëtë trouvës in well-enriched corners of housing plots.

4.4.1.7. Distribution of fruit trees

The distribution of fruit trees is given in Table 28. For reasons of space it will be coupë into several sub-tables a, b and c.

Table 28a. Rëpartition of fruit trees diffèrent from breadfruit in the plots.

Meng that have at least one tree	Meng who have at least one espëce	Number of plants of this species	Number of households with two species	Number of plants of this species	Number of households with three species	Number of feet of these species	Number of households with four species	Number of feet of these species
15	3	7	1	2	0	10	4	17
12	3	3	2	5	0	5	4	20
8	0	0	1	2	0	13	5	14

2	0	0	0	0	0	5	0	0
45	3	7	11	36	0	60	8	38
57	13	25	7	24	0	40	7	18
7	0	0	1	2	0	10	4	7
34	4	5	12	27	1	30	4	23
Tot. 180	31	47	35	101	1	259	33	228
%43 ,5								
100	17,3		10,5		0,5		18,4	

Lëgende: Tot. = Total; Мёпд = mënages.

From Table 28a, it can be seen that 0.5% of the mënages have three tree espëces with a total of 259 tree feet, 18% have four with 228 individuals. In addition 17% have one espëce with 47 individuals and 10% have two espëces with 101 feet.

Table 28b shows the distribution of fruit trees other than breadfruit trees located outside the plot.

Table 28b. Rëpartition of fruit trees diffërent from 1 breadfruit tree outside the housing plot.

No. of mg. who have at least one amëliorë tree	No. of mg. who have less than one species	No. of plants of this species	No. of mg. with two species	No. of feet of these species	No. of mg. with three species	No. of feet of these species	Number of mg. with four species	No. of feet of these species
2	0	0	0	0	0	0	1	1
1	0	0	0	0	0	0	0	0
0	0	0	0	0	0	0	0	0
1	0	0	0	0	0	0	0	0
2	2	2	1	2	0	0	1	5
0	3	3	3	18	0	0	0	0
0	0	0	0	0	0	0	0	0
1	0	0	1	3	1	3	0	0
Tot. 7	5	5	5	23	1	3	3	22
%3 ,9	2,7		2,7		0,5		1,6	

Lëgende : Tot = Total ; Nb. = Number ; Mg. = mënages.

Table 28b shows that 3.9% of households have at least one improved tree, i.e. a total of 7 feet. 2.7% have a different species of tree from the breadfruit tree outside the housing plot, with 5 feet: 2.7% have two, with 23 feet: 0.5% have 3 feet outside the plot and 1.6% have four, with 22 feet.

Table 28c shows the number of households with at least one breadfruit or other tree.

Table 28c. Number of mënages with at least one breadfruit or other tree.

Number of households with at least one	Number of feet	Number of households with at least one	Number of feet	Number of households with another	Number of feet	Number of households with at least one	Number of feet

apain tree in the plot		apain tree outside the plot		type of tree on the plot		other type of tree outside the plot	
0	0	0	0	0	0	0	0
1	1	0	0	0	0	0	0
0	0	0	0	0	0	0	0
0	0	0	0	0	0	0	0
1	1	0	0	1	1	0	0
2	0	0	0	0	0	0	0
0	0	0	0	0	0	0	0
1	1	0	0	0	0	0	0
Tot. 5	5	0	0	1	1	0	0
%2 ,7		0,0		0,8		0,0	

Lëgende : Tot. = Total.

Analysis of Table 28c shows that 2.7% mënages have at least one breadfruit tree with 5 feet and 0.8% have at least one other type with 1 foot in the plot.

1.1.1.6.   Number of fruit trees and other fruit-bearing trees per species planted per household.

Table 29 shows the number of fruit trees (and other trees) per species planted per household. For reasons of space table 29 will be divided into a, b and c.

Table 29a. Number of feet of fruit trees (and other) by species and тёпаде.

Number of households with at least one fruit tree	Number of mango trees	Number of Avocado tree	Number of plum trees	Number of citrus plants	Number of Safoutier vines	Number of Pomier plants
15	8	8	1	5	1	3
11	8	8	0	2	3	2
8	8	6	1	1	2	1
2	2	1	0	0	0	0
44	37	30	4	11	10	2
52	33	21	0	14	10	5
7	4	3	0	0	1	1
33	93	25	10	1	3	3
Tot. 172	125	87	7	36	30	19

Lëgende : Tot. = Total.

Table 29a shows that the survey identified 125 mango trees in 172 households surveyed. Mango remains the most represented species. It is followed by avocado (87 trees) and citrus (36 trees).

Figure 4 shows a photo of a safoutier (*Dacryodes edulis*) on the outskirts of the town.

Figure 4. Safoutier plants in residential plots in the vicinity of the town.

As you can see, the people of Kinshasa, whether they live in the city centre or in the outskirts, always plant a fruit tree in their homes for food security.

Table 29b. Number of feet of fruit trees (and other) by espëce and mënage (continued).

Number of households with at least one tree	Households with a coconut	Households with papaya	Menage quiont the badamier	Households with the heart of beef	Guava-growing households	Households with the carambolier
15	2	3	1	2	1	1
11	1	1	0	2	0	0
8	0	4	0	2	0	0
2	0	2	0	0	0	0
44	2	23	1	6	2	0
52	2	9	2	4	1	0
1	2	3	1	1	0	0
33	2	14	1	4	1	0
Tot. 172	11	59	6	21	5	1

Lëgende : Tot. = Total.

Analysis of this table shows that papaya (with 59 plants) comes after the two already mentioned (mango and avocado) planted by mënages. It is followed by creur de breuf.

Table 29c shows the number of breadfruit trees and other fruit species.

Table 29c. Number of feet of fruit trees (and other) by espëce and mënage (continued).

Number of households with at least one tree	Households with breadfruit	Households with palm trees	Households with banana tree	Households with cane sugar
15	0	5	1	0
11	1	3	3	0
8	0	1	3	0
2	0	0	2	0
44	1	17	9	0

52	2	15	8	4
7	0	4	5	0
38	1	14	8	1
Tot. 172	5	59	39	5

Lëgende : Tot. = Total.

Analysis of table 29c shows that oil palm (*Elaeis guineensis*) (59 feet) first and banana *(Musa paradisiaca)* (59 feet) second are widely cultivated by households in Limete.

Mango, avocado, papaya, palm, banana and citrus trees are the major species found on housing plots in this Commune.

1.1.1.7.   Animal species raised by households

Table 30. shows the animal species raised by the households. For reasons of space, this table will be divided into a and b.

Table 30a Animal species reared by households.

Breeding households	Hen	Duck	Pigeon	Guinea fowl	Guinea pig	Pork
6	4	0	1	1	0	0
3	2	0	0	0	0	0
3	2	1	0	0	0	0
0	0	0	0	0	0	0
14	9	0	1	0	1	0
19	13	0	0	0	0	1
2	1	0	0	0	0	0
18	12	4	0	0	0	0
Tot. 65	43	5	2	1	1	1
%100	66,1	7,6	3,0	1,5	1,5	1,5

Lëgende : Tot. = Total.

Analysis of this table shows that 66, 1% of mënages ëlëyen1: at least one hen and 7.6% a duck.

Figure 5 shows a photo of a chicken farm.

Figure 5. A free-range hen.

This photo shows that in more than one case the hens are ëlevëes roaming.

Table 30b shows the animal species reared by households.

Table 30b. Animal species reared by households (continued).

Breeding households	Goat	Chicken and duck	Gable and duck	Hen and pigeon	Chicken and pork	Poulean d guinea pig
6	0	0	0	0	0	0
3	0	1	0	0	0	0
3	0	0	0	0	0	0
0	0	0	0	0	0	0
14	0	0	0	1	1	0
19	(1)	2	1	1	0	1
2	0	0	0	0	0	0
18	0	2	0	0	0	0
Tot. 65	(1)	6	2	2	1	1
%100	1,5	9,2	3,0	3,0	1,5	1,5

Lëgende : Tot. = Total. () = not seen by interviewer.

This table (30b) shows that hens and ducks are more closely associated in Kinshasa. Also very few of the mënages ëlëvent the animals of two different speciës simultaneously (hen and pig or hen and guinea pig).

1.1.1.8.   Number of heads per species bred

Table 31 shows the number of livestock-raising households and head per species. For reasons of space this table will be соирё in table a and b.

Table 31a. Number of mënages breeders and heads ëlevëes by espëce.

Number of farms that raise hens	Number of hen heads ëlevëe	Number of farms raising ducks	Number of duck heads ë^ë	Number of farms raising pigeons	Number of sprocket heads ë^ë	Number of farms raising guinea fowl	Number of guinea fowl heads ë^ë
4	26	0	0	1	4	1	2
3	17	1	1	0	0	0	0
2	4	1	2	0	0	0	0
0	0	0	0	0	0	0	0
11	29	1	7	1	11	01	0
17	42	3	8	2	4	0	0
2	4	1	2	0	0	0	0
14	52	6	15	0	0	0	0
Tot. 53	224	13	35	6	19	1	2

Lëgende : Lëgende : Tot. = Total ; ( ) = not viewable by enqueteur.

This table shows that the hen is the most ëlevëe species (recensëe). It is found in the most mënages and has the most ëlevëes heads. It is followed by duck in number of mënages practising its ëlevage and also in number of ëlevëes heads.

Table 31b shows the number of households practising livestock farming and the number of head per species.

Table 31b. Number of breeding mënages and heads ëlevëes per espëce (continued).

48

Number of households driving a guinea pig survey	Number of guinea pig heads bred	Number of households raising pigs	Number of pig heads bred	Number of households raising goats	Number of bred goat
0	0	0	0	0	0
0	0	0	0	0	0
0	0	0	0	0	0
0	0	0	0	0	0
1	8	1	8	0	0
1	3	1	2	2	(8)
0	0	0	0	0	0
0	0	0	0	0	0
Tot. 2	11	2	10	2	(8)

Lëgende : Tot. = Total ; ( ) = What is not seen by the interviewer.

This table shows that the number of households rearing guinea pigs, pigs and goats is the same, only the number of heads reared is different. The number of guinea pig heads is higher than the number of pig heads. Goat rearing is conducted outside the housing plot and its surroundings.

1.1.1.9. Breed specificity

Table 32 shows the spëcificitë of the hen breeds (either village or aтëHorëe).

Table 32. Schicken breeds.

Number of hen	Village chicken	Laying hens	Poulede flesh	Mixed hen (chair and layer)
4	2	2	0	0
3	2	1	0	0
2	2	0	0	0
0	0	0	0	0
11	11	0	0	0
17	16	1	0	0
2	1	1	0	0
14	12	2	0	0
Tot. 53	46	7	0	0
%100	86,7	13,2	0,0	0,0

Lëgende : Tot. = Total.

Table 32 shows that indigëne (village) hens are the most common (86.7%) on these farms. No amëliorëe (broiler and mixed breed) hens were ëtë recorded (0.0%).

4.4.2. Agricultural production

4.4.2.1. Estimated vegetable, fruit and meat production in the residential plots and surrounding area

Tegume production (actual and potential) and fruit production are shown in Table 33.

Table 33a. Production of k'gume (in tonnes) and fruit (in kg).

Production	Production	Producti	Production	Production	Producti

current vegetable	possible vegetable	wave mango tree	from avocado	palm tree	ondu papaya
0,213	0,573	800	640	437,5	258
0,276	0,594	800	640	262,5	86
0,77	0,327	800	480	97,5	344
0,100	0,142	200	80	0,0	172
1,227	4,821	3600	2400	1487,5	1978
0,931	3,323	3300	1680	1312,5	774
0,861	0,951	400	240	350,0	258
1,074	5,904	2500	800	1225,0	1204
Tot. 4,159	16,635	12400	6960	5152,5	5074
Production / tonne per day the survey 4,155	16,635				
Production per species/ tonne/year 33,272	133,08	12,4	16,96	5,1625	5,07

Lëgende : Tot. = Total.

Analysis of this part of the table (33 a) shows that current vegetable production (33 tonnes) is a quarter of what could be produced if the space available on these plots were exploited (possible production of 133 tonnes) per year. This was even the case on the day of the survey (4 tonnes compared with 15.5 tonnes).

It also shows that mango production is almost double that of avocado, palm and papaya. Banana production is the lowest.

Table 33b shows the second part of the production table. It groups together data on fruit and meat production.

Table 33b. Total fruit and meat production by species.

Prod. total fruit	Prod. dela hen	Duck production	Pigeon production	Guinea pig production	Pork production	Prod. de the goat	Prod. total meat (Kg)
3,0305	35,1	0,00	1,92	0,00	0,00	0,00	37,02
2,1985	22,9	3,15	0,00	0,00	0,00	0,00	26,05
1,9465	5,4	6,30	0,00	0,00	0,00	0,00	11,70
0,4920	0,00	0,00	0,00	0,00	0,00	0,00	0,00
11,5705	106,0	22,05	5,28	8,00	2600,0	0,00	2741,33
9,6765	56,7	25,20	1,92	3,00	550,00	420,00	1156,82
1,3480	5,4	6,30	0,00	0,00	0,00	0,00	11, 70
6,4140	70,2	47,25	0,00	0,00	0,00	0,00	117,45
Total 36,7565	302,4	110,25	9,12	11,00	3200,0 0	420,00	4102,07
Production	0,3024	0,1103	0,009	0,01	3,250	0,420	4,1027

in t/day of survey							

Lëgende : Prod. = Production.

Table 33b shows that total fruit production is 36.7 tonnes per year. Meat production is 4.1 tonnes, with pork production at 3.2 tonnes at the time of the survey.

At the time of the survey, only pork was producing 3 tonnes.

This production is interesting when measured in kilograms (on adult live weight).

Figure 6 shows a ĕ pig farm being run in a masonry-built enclot.

Figure 6. Plot-based pig farming in Kinshasa.

The pigs are reared in pens built either of wood or masonry.

For fruit production, Table 34a shows the quantity of fruit and daily availability of fruit per inhabitant of the Commune for the 6 major species. For reasons of space, this table will be divided into a and b.

Table 34a. Quantity of fruit and daily availability of fruit per inhabitant of the Commune for the 6 major species.

Species	Number of feetat production per species extracted from all 18,475 plots in the Commune	Production at tonnes extrapolated to all the trees in the Municipality	Average food availability per per person per day inhabitants of the Communein gramme extrapolated across all plots	Food availability per personand per day, the Commune's residents inKcal extrapolated across all plots
Mangifera indica	11.488	1.146, 789	10,9	4,54
Persea americana	7.982	638,532	6,1	5,82
Elaeis guineensis	5.413	473,670	4,5	14,58

Carica papaya	5.413	485.505	4,4	1,04
Musa paradisiaca	3.394	67.890	0,6	0,34
Dacryodes edulis	2.752	192,661	1,8	2,37

This table shows that *Mangifera indica* is the species that is most represented in the commune. But it is the food availability of *Elaeis guineensis* that is important.

Those (quantile and daily availability) for minor espëces are shown in table 34b.

Table 34b. Quantity of fruit and daily availability of fruit per inhabitant of the Commune for the 13 minor species.

Espëces	Number of feet in production parespëce extrapolated across the18 ,475 plots of land owned by the Commune	Production tonnes desarbres inventories in the sample	Production tonnes extrapolated to all trees dela Municipality	Food availability per person per day inhabitants of the Commune in grams in the plots who have these espëces	Food availability per person per day inhabitants of the Commune in Kcal in plots that have these species
Annona reticulata	1.9227	0,153	28905.	2,63	1,17
Citrus limon	1.743	1,349	123,761	12,46	1,30
Eugenia malaccensis	1.743	1,330	122,018	-	-
Citrus sinensis	1.193	0,832	76.330	11,23	2,43
Terminalia catappa	1.101	0,972	34,728	-	-
Cocos nucifera	1.009	5.784	52,807	91,96	227,23
Flacourtia ramoutchi	459	1,500	137,615	52,85	22,38
Artocarpus incisa	459	0,095	8,716	3,33	1,94
Psidium guajava	183	0,194	17,798	17,02	5,88
Citrus reticulata	92	0,020	1,835	3,51	2,68
Musa sapientum	92	0,007	0,642	1,23	0,35
Passiflora edulis	92	0,019	1,743	3,33	-

Table 34b shows that *Cocos nucifera* is the species with the greatest availability on the plots where it is available.

4.4.2.2.   Main destination of household production (of vegetables, fruit and meat in the home plot and surrounding area)

This part of the results will be divided into three tables: 35a for the vegetable garden, 35b for fruit trees and 35c for livestock.

Table 35a. Main use of vegetable garden produce.

Mënages with gardens	Self-consumption	Sale	Self-consumption and sale
9	7	0	2
11	10	0	1
5	3	0	2
2	2	0	0
39	30	0	9
33	28	1	4
7	5	0	2
31	24	0	7
Total 137	109	1	27
% 100	79,5	0,7	19,7

Table 35a shows that over 79% of households with vegetable gardens use their produce for self-consumption. Only 0.7% of households sell their produce.

Table 35b shows the main destination of fruit tree production.

Table 35b. Main destination of fruit tree production.

Mënages with fruit trees	Self-consumption	Sale	Self-consumption and sales
15	13	0	2
11	8	0	3
8	7	0	1
2	1	0	1
44	30	0	14
52	40	0	12
7	4	0	3
33	20	0	13
Total 172	123	0	49
% 100	68,7	0,0	31,2

Table 35b shows that 68.7% of households use their fruit production for self-consumption and 31.2% for sale and self-consumption.

Table 35c shows the destination of livestock production.

Table 35c. Main destination of livestock production.

Mënages who have a ëlevage	Self-consumption	Sale	Self-consumption and sales
6	5	0	1
3	2	0	1
3	3	0	0
0	0	0	0
14	10	0	4
19	16	1	2

2	1	0	1
18	17	0	1
Total 65	54	1	10
% 100	83,0	1,5	15,3

This table shows that 83 of the households intend to use their livestock production for own consumption. 15.3% use it for self-consumption and sale and 1.5% for sale.

4.4.2.3. Farmers' main motivations

The motivations behind these farmers are grouped together in table 35.

Table 36. Farmers' main motivations.

Number of households practising agriculture	Number of farmers	Childhood habits	Reduce expenditure on the household budget	Increasing household income	Stimute spoke agricultural training	Rë inconsistent answer
15	19	3	3	2	3	8
12	11	2	3	0	0	4
8	8	4	1	1	1	1
2	2	0	0	0	1	1
46	38	7	10	6	9	4
60	125	15	75	20	5	10
7	9	2	5	1	1	0
38	41	7	18	7	4	5
Total 188	253	40	117	39	24	33
%	100	15,8	46,2	15,4	9,4	13,0

Table 36 shows that 61% of farmers became involved for economic reasons, i.e. 46% to reduce household budget expenditure and 15% to increase their тёпаде income. Only 13% of farmers put forward very confused motivations but, in rëaШё, they hide the difficulties due to the current situation.

4.4.3. Rice, vegetable, fruit and meat production at off-plot sites

Off-plot production is groupëe in Table 37. For reasons of space, this table will be всирё in a, b.

Table 37a. Agricultural production oaysanne des mënages.

Sitede production	Current surface area (ha)	Possible surface area (ha)	Current production (Kg)	Possible production (Kg)	Current total production (tonnes)	Total possible production (tonnes)
Kingabwa rice mill	400,00	520,00	1, 500de paddy	6,000 Kg of paddy	600de paddy 400 of husked riceë	3121de paddy 2185de husked rice
Limete	0.10of which	0.12of which	3000de	6000de	0. 09de	0. 3de

Agricultural Farm	0.0325 of enriziere fish farm 0. 042de fish farming 0. 023de lklevage	0.05 of enriziere fish farm 0.05 of fish farming 0.025 dklevage	paddy 100 of poison A few kilos of meat (rabbit, duck, pork)	paddy 200de fish Some meat (rabbit, duck, pork)	paddy 0.06 of husked rice 0. 01 of rice fish and several hundred kilos of meat	paddy 0.21 of husked rice 0. 02 of rice fish A few hundred Kg of meat
Market garden Limete interchange	0,7	0,7	1000de kgume en 3 to 4 months	1000de kgume in 3 to 4 months	2.1 of kgume	2.1 of kgume
Mombele Agricultural Technical Institute	0,2	0,7	1000de kgume en 3 to 4 months	7000de kgume in 3 to 4 months	6of kgume	21of kgume
Centre d'Accueil etde Passage	0,05	0,05	60de cassava leaf in 1 to 2 months	60de cassava leaf in 1 to 2 months	0. 3of cassava leaves	0. 3of cassava leaves
Lelong from	0,5	0,5	2500 sheets of	2500 sheets of	15 of sheets of	15 of sheets of
avenues Poids Lourd, Universite and Boulevard Lumumba			manioc	manioc	manioc	manioc
Total	401.64	522,39			600.09 of paddy 420.06 of husked rice 0. 61de fish A few hundred Kg of meat 27de vegetable 15. 3de cassava leaves	3121.55 de paddy 2185.55 deriz decortic 0. 02de fish A few hundred kilos of meat 42of vegetable 15. 3de cassava leaves

Analysis of Table 37a shows a potential area of 522 ha that could be used for agricultural activities, of which only 401 ha have been cultivated to date. A production of 600 tonnes of paddy, 420 tonnes of hulled rice, 0.61 tonnes of fish, 27 tonnes of vegetables, 15 tonnes of cassava leaves and several hundred kilograms of meat are obtained. If the entire plot were exploited, production would reach more than 3121 tonnes of paddy, 2185 tonnes of hulled rice, 0.02 tonnes of fish, 42 tonnes of vegetables, 15 tonnes of cassava leaves and several hundred kilograms of meat.

Figure 7 is a photo of a matembele *(Ipomoea batatas)* leaf garden with a papaya plant.

Figure 7. Matembele garden, which also bears a papaya tree (*Carica papaya*) in a production site overrun by unplanned buildings.

Table 37b groups together the output of structures spëcialisëes in professional trade.

Table 37b. Production of spëcialisëes structures.

Transformationstructuresand broodstock production	Current production
Kimbanguist Centre (1995)	460 broodstock (chicks) + around 2,000 eggs covered
CDI DevelopmentCentre Bwamanda (1995)	11,830　　　　　children 6,576　children 131,335 reufs fecondes on 3,849 t. of animal feed 78. 73%sponge food 9.　07%fleshfood 6.　04%porkfood 2.34% horse　feed 2.04% quail feed
Soyapro ONG (1995)	Average5kg/d of　　　fungus 42,125 bottles of soya milk 49,154 sachets of soya milk 316 bottles of various wines
SocieteAgricoleetVeterinaire (SAVET) (　　lastproduction anterior)	With 40,000 layers, it produced 1,200-1,220 trays of 30 eggs per day = 3,600 eggs for consumption.

	10 to 11t. of feed

Table 37b shows that the processing and broodstock production structures (notably chicks) produced more or less 2,000 eggs covered by the Centre Kimbanguiste. The Centre de Dëveloppement Integre (CDI)-Bwamanda, Soyapro-ONG and the Socïëtë Agricole et Veterinaire-SAVET contribute greatly to the production of food for humans and livestock.

4.5. New developments

- The city is experiencing significant population growth. Estimated at 6.062 million in 2000 (INS, 1993), the population is expected to reach 10 million in 2020 (Makumbelo et al., 2023). Some sources even estimate that this population will oscillate in 2023 between 15.628 and 16.316 million inhabitants in the city, with a density of 27,193 inh./km^2 ; while for l'аддlотёгайоп of Kinshasa, it is estimated at 17.239,463 million inhabitants with a density of 1,730 inh./km2 in 2021 (Congo Job ecer.com, 2023; en.m.wikipedia.org).

- A change in the heads of households and the disappearance of vegetable gardens in the plots where they were once located.

- The habit of planting vegetables: cassava leaves (*Manihot glaziovii*) and matembele (*Ipomea batatas*) and fruit trees: avocado (*Persea americana*), papaya (*Carica papaya*), mango (*Mangifera indica*) and safoutier (*Dacryodes edulis*) close to the household.

In 2011, a survey on urban agriculture in Kinshasa in a sample of 500 market gardeners identified individuals of *Mangifera indica* and *Persea americana* with a frequency of 200 and 58 respectively in all the plots surveyed and recorded the opinions of 97,4% of those surveyed said they "enjoyed market gardening" and 97.0% said they were "satisfied with market gardening because its income allows them to live" (Musibono et al., 2011).

Figure 8 shows a piecemeal plot where more than one fruit species is cultivated or maintained for the sëcuritë alimentaire du mënage.

Figure 8. A piecemeal plot bearing more than one fruit species.

Analysis of this figure shows a half-cut plot in which a plant of *Dacryodes edulis* (*Burseraceae*), *Carica papaya* (*Caricaceae*), *Mangifera indica* (*Anacardiaceae*), *Persea americana* (*Lauraceae*) and *Spondias dulcis* Foster (syn. *Spondias cytherea* Sonn. (manga ya sende) (*Anacardiaceae*) and in the background two other individuals of *Persea americana* planted or maintained for the household's food supply.

-     The modernisation of the town has prompted many residents to build on the upper floors or to cement their plot of land.

It has already been mentioned that when the population of a town increases, it gives way to an increase in built-up areas and a loss of space for growing food and competing.

-     While in the first decade, the surface area of a plot was 4ares 4ca (law n°80-008 of 18 July 1980), by the end of this decade, more than one housing plot had been split up due to the increase in the number of residents, the modernisation of the town and the increase in space for other lucrative activities. This has seriously reduced the space available for gardens and livestock on housing plots.

-     The need for housing land has increased drastically, leading to the subdivision of former vegetable production sites. Some have lost significant areas of land and others have even disappeared.

-     Some trees, notably mango (*Mangifera indica*), safoutier (*Dacryodes edulis*) and coconut (*Cocos nucifera*), are very tall.

-     The tendency to occupy all the land has led churches, schools and the public administration to close their concessions as a precautionary measure.

-     Terraces are built along major roads because of rainwater and fear of erosion.

-     Households' need for food has increased. As a result, the sale of vegetables, fruit and meat has become a highly profitable business.

-     This urban sprawl has prompted a number of individuals and/or legal entities with substantial financial resources to transfer the pool of agricultural activities to the outskirts of the town, to the Plateau des Bateke in particular.

-     On the outskirts of Kinshasa, large tracts of land are occupied by agricultural farms belonging to private nationals and/or expatriates.

-     This results in conflicts between tribes and between "indigenous and non-indigenous populations".

The two figures (9 and 10) show photos of former agricultural production sites now occupied by buildings.

Figure 9. Photo of the concrete buildings where the Matete market garden was once located.

An analysis of this figure shows that the buildings that have sprung up occupy the Matete production site, where market gardeners used to produce tonnes of vegetables a year, a few decades later.

Figure 10. Photo of l'Echangeur de Limete where the production site of l'Echangeur was located.

Analysis of this photo shows a totally fenced-in site where everything is under bëton with leisure areas on all sides.

Chapter V. INTERPRETATION OF THE RESULTS AND ECODEVELOPMENT PERSPECTIVES FOR 2025-2035

5.1. Interpretation of results

5.1.1. Nutritional status

Women are aware that the most important factor in ensuring their children's growth is good nutrition (42.5%).

In the field, unfortunately, the percentages of malnourished children are higher than the normative rates of 10%, 20% and 15%. The actual situation is as follows: iiguc' malnutrition: 11.4% of children are malnourished according to the P/T ratio and 12.3% of children have a brachial përimëtre of severely malnourished children. According to the same criterion, 17.5% of children were moderately malnourished. 36.0% of children had global malnutrition, according to the P/A ratio, and 13.1% had chronic malnutrition, according to the T/A ratio. In addition, a significant number of children are underweight in maternity wards, i.e. 5% in Mososo and 16% in Kingabwa 1 and 2. It is accepted that if "in дёгагя! the percentage of malnourished children is equal to or greater than 10% using ^ёra^ the brachial përimëtre or the P/T ratio, malnutrition is widespread and there is a real nutrition problem. On the other hand, if, using the T/A ratio, a rate greater than 20% is found, there is a major problem, as is the case when the P/A ratio is 15%" (CEPLANUT, 1996).

This major nutritional problem explains the deteriorating health of the population in this Commune. Curiously, it is the Kingabwa pecheur, Maman Nzenze and Funa neighbourhoods that are most affected. And yet these neighbourhoods are inhabited by fishermen.

5.1.2. Food safety

The difficulties faced by households with regard to food security in Kinshasa have been highlighted in more than one publication. Maisonneuse (1999) is a case in point.

The results of this ëtude allow an interpretation according to which :

48.2% of farmers are aware that with good permanent agricultural production, a household can ensure food security for its members. 32.0% thought they could do so with financial means alone, 17.3% by producing and selling the products of certain income-generating activities (sewing and selling clothes, making and selling cakes, etc.).

As for the actual availability of food, the agricultural production of those who cultivate, plant and keep livestock outside the plots (Valtee de N'djili, Ferme agricole de Limete, Echangeur de Limete, Institut Technique Agricole de Mombele, Centre d'Accueil et de Passage Kimbanguiste and along major roads) can surely ensure that their households have access to vegetables, fruit and livestock products for more than 3 days a week. Production along major roads raises fears of lead-related diseases (Monama et al., 1985).

Only 15.9%, 12.8% and 0.1% of households have vegetables, fruit and livestock products available more than 3 days a week, respectively.

Cassava and maize (18.4% of households), cassava (7% of households), cassava, chikwangue and malemba (3.1% of households) are the staple foods available for more than 3 days a week.

Very few households enjoy food security in the commune. "The food situation is critical. The thorny problem has two aspects: unstable supplies and low purchasing power" (CEPLANUT, 1996). This situation is at the root of what the FAO had already stated in 1970 for the country as a whole: "a large proportion of the population suffers from serious calorie and protein deficiencies" (Ministëre de l'Agriculture, Animation Rurale et Dëveloppement Communautaire, 1991). However, "the corn once ignored by the population of Kinshasa is rapidly being adopted by that population" (Vangu, 1995).

5.1.3. Urban agriculture

The off-plot imiraieliere production sites have sufficient space and currently produce a significant amount of vegetables. This production could be increased if these areas were used wisely. However, it should be pointed out that production on some sites is already polluted by one of the great rivers of modern civilisation. Lead levels exceed WHO standards (not exceeding 10 ppm in foodstuffs) (Monama et al., 1985).

Only gardens and fields in urban environments, outside forest and savannah ecosystems, can provide vegetables for people who do not have sufficient financial means.

As can be seen, the most widely grown vegetables are those that do not require a nursery or transplanting. *Manihot glagiovii*, *Phytolacca dodecandra* and *Psorocarpus scandens* (fresh weight per 100 g of edible part) have leaf protein contents of the order of 6% to 7% and exclusive fibre carbohydrate contents of 5% to 7%, while *Justicia strata* subsp insularis, which has been recently cultivated, has contents of 4% and 9% respectively (Dupriez and Liener, 1987; Mbemba and Remacle, 1992). *Ipomea batatas* has no less.

All in all, our results showed that 19 vegetable species were cultivated and around 47,000 plants of the 18 fruit plant species were planted or maintained on 1.09% of the 18,475 plots in this Commune. *Mangifera indica* was the most frequently planted tree species and *Ipomoae batatas* the most frequently cultivated vegetable species.

Households in the Commune's housing plots contributed 33.27 tonnes of vegetables, 4,087 tonnes of fruit and 4.1 tonnes of meat per year. Vegetable production could reach a possible 133.08 tonnes if the entire available area were exploited.

The six major species (*Mangifera indica*, *Persea americana*, *Elaeis guineensis*, *Carica papaya*, *Dacryodes edulis* and *Musa paradisiaca*) and citrus fruits alone produced 36.67 tonnes of fruit. The contribution of these species to the population's diet was estimated at 10.9g, 6.1g, 4.5g, 4.4g, 1.8g and 0.6g of fruit per person per day respectively. This equates to an average daily food availability for the inhabitants of the commune for all the plots with these species of 4.54 Kcal; 5.82 Kcal; 14.58 Kcal; 1.04 Kcal; 2.37 Kcal and 0.34 Kcal. The mastery of certain ecotechniques and environmental education relating to 'trees in the city' could increase the importance of this contribution. These results have also been published by Makumblo et al (2002, 2005).

A survey of *Burseraceae* in 12,977 plots in 10 districts of 5 Communes of Kinshasa showed that 599 plots, or 5% of plots, had at least one *Dacryodes edulis* plant (Makumbelo et al., 2016).

The Commission Interministerielle (1998) estimates that there is one plant of *Mangifera indica* and one plant of *Persea americana* in each plot, and one plant of *Dacryodes edulis* in one plot out of 3 in Kinshasa. This is in contrast to Limete where individuals of these 3 species were found in 50%, 42% and 14% of the plots surveyed respectively.

Fruit arboriculture in market gardens and on housing plots is one of the activities that helps to increase the availability of food for households in Kinshasa (Ministries: Planning, Agriculture and Livestock, National Education, Environment, Nature Conservation and Tourism/UNDP/UNOPS, 1998, quoted by Makumbelo et al., 2005). At the same time, it makes it possible to reconcile the conservation and sustainable use of biological diversity in its immediate environment with the development of the population. This is the concern of the IUCN (International Union for Conservation of Nature and Natural Resources, 1980), Our Common Future (World Commission on Environment and Development, 1988) and the Rio

Conference (United Nations Conference on Environment and Development - UNCED, 1993), as taken up by Makumbelo et al. (2005).

A large number of tropical vegetables and fruits have a particularly high nutritional value. Their significant contribution in calories, proteins, various minerals useful to the body and vitamins (Dagroote, 1970; FAO, 1990; Mbemba and Remacle, 1992; Favier et al., 1993) is an important asset for human nutrition and health.

The major species (*Mangifera indica*, *Persea americana*, *Elais guineensis*, *Carica papaya* and *Dacryodes edulis*) are found among the 29 fruit species cited by Pauwels (1982, 1993) as common or less common trees in modern and customary environments in the Kinshasa region, along with 8 others that he mentions as minor species. The first species has (by fresh weight per 100 g of edible part) a calorie content of 65 Kcal, a protein content of 0.6 g and an exclusive fibre carbohydrate content of 17.2 g; the second has 159 Kcal, 1.8 g and 8.0 g respectively; *Dacryodes edulis* has 263 Kcal, 4.6 g and 14.9 g (Mbemba and Remacle, 1992). This urban vëgëtation fulfils a number of functions, in particular food,

medicinal, artistic and magico-religious (Kabeya et al., 1994; Makumbelo et al., 2016). It also helps to maintain the biogëoclimatic cycle in urban ëcosystëmes. Reinforced, it can faggot the biodiversity and physiognomy of the urban ëcosystëme, allow the reduction of substances and dust in the atmosphëre and stabilise its composition and climatic effects. This role of thermal regulation is only possible when the presence of trees and vegetables is evident in all the plots, streets and other areas of the town and city.

Unfortunately, the demographic boom and the modernisation of the city have led to urban sprawl and its consequences, which Guillaume Faburel deplores. He points out that when a city's population increases, it loses its cultivated areas, wetlands and plant cover, and gives way to an increase in built-up areas, which can rise by as much as 134%. This is the case in Dhaka and Mumbai, where the built-up area has almost doubled (Faburel, 2023). This results in direct or indirect competition between members of different households.

5.2. Ecodevelopmental outlook for 2025-2035

If the 1990-2000 decade was marked by the destruction of the economic and health situation of the population of Kinshasa, caused essentially by the refusal to implement the resolutions of the Sovereign National Conference, the dislocation of access routes to the capital and the effects of the war of liberation led by the AFDL and others, new developments are emerging as new problems for Kinshasa in the 20252035 period. What we need to think about is another ecodevelopmental scenario to answer the central question: "how can we ensure food security for the entire population in households, or how can we encourage households to feed themselves properly (rather than) how can we farm or raise livestock?" (Sachs et al., 1981).

The solution to this question may lie in an adapted urban agricultural response in which the garden of residential plots integrates soilless cultivation (Aya, 2023) and agricultural activities outside residential plots on farms supported by an institutional framework and a national production fund. Revised fruit arboriculture, for which the ecotechniques of propagation, improved production, processing and conservation would merit more attention from farmers. Research will focus on multiplying improved or selected varieties, updating the quantification of production by species and determining the best edaphic conditions. Environmental education on 'trees in the city' will draw people's attention to the importance of vegetables and fruit trees in urban ecosystems and in urban agriculture (Makumbelo et al., 2005).

Back in 2011, it was suggested that the governing state should invest in securing market gardening economically by granting microcredits, ecologically by drastically reducing the use of chemicals and socially by promoting the propriëtë Гопаёге, because the market gardeners dëclarent that market gardening allows them to have money and to live better than the civil servant, to send the children a lkcole and to live on 9a for a month (Musibono et al., 2011). Soilless cultivation is a type of agriculture conceived of simultaneously by two Germans, KNOP and Sachs, in 1860. Its real development dates from the years 1975-1980. The advantages of this type of cultivation include the use of areas where there is no soil, building terraces and rubbish tips (Morard, 1995). It is therefore one of the types of agricultural practices adapted to the conditions of urban environments that emerge from the soil.

CONCLUSION

This work focused on the problems of food and nutrition in urban areas. The results of the surveys enabled us to draw the following conclusions: > a large number of urban farmers produce various foods throughout the commune; > the production of these farmers can contribute to the food security of the people of the Commune of Limete in particular and of Kinshasa in general, shaken by the effects of a multiform crisis, but that very unfortunately, the health of the population continues to deteriorate, the following was affirmed:

1° The lack of food available in households contributes to the poor health of the   population of the city of Kinshasa;

2° some of the foodstuffs (vegetables, mushrooms, fruit, seeds, fish, meat, etc.) consumed in Kinshasa are produced (or picked) locally; 3° the majority of those who practise urban agriculture show very little interest in concerted and sustained action among themselves, even though they are often financially   deprived and under-equipped households   ;

4° generally, they practise the usual techniques, without any effort at improvement; 5° good coordination of all the structures devoted to urban agriculture would be an asset for improving these activities with a view to making a lasting contribution to the food security of household members in the city of Kinshasa. The case of the Commune of Limete.

In order to verify this hypothesis, a study was carried out in the Commune of Limete to find answers to the question of how to ensure food security for the household population, or how to encourage households to feed themselves properly (rather than how to teach them to farm or raise livestock), with a view to examining the possibilities of finding a solution for the health of the town's population.

The results of this latest study showed :

- insufficient effort is being made to develop the various food resources that grow spontaneously to supplement the range of foods already cultivated in the capital;
- the virtual absence of a social production system. Farmers work individually. As a result, it is difficult to supervise and support them in the implementation of any recovery plan. Also, they have almost no access to any support in the case of smallholders. They are scattered throughout the town, cultivating and/or raising livestock on small areas of land that are not sufficiently recognised and protected by the public authorities;
- I insuffisance de rationalite et de l'efficacite des ecotechniques adoptёes. The most adaptёes can produce more food of good quality and in a гёдиНёге way, while caring for future generations with respect for the environment. They use the usual mёthodes without mastering the consёquences of their actions. This explains the meagre harvests, which are unquantified and even unquantified in figures, the presence of multiple forms of contamination via fruit, vegetables and meat, and the impoverishment of the soil. Alongside these shortcomings, there are also those of appropriate techniques for preserving and processing this production;
- food insecurity, both in terms of plant and animal food resources and in terms of the staple foods consumed in the town. This is one of the reasons for the poor health of the Commune's population;
- the lack of credit facilities or other sources of finance for small and large-scale operators (of production and processing sites);
- the insecurity created by looting and political quarrels in the country.

As a result, a number of improvements have been envisaged and suggested with a view to setting up an institutional framework capable of popularising and defending the developmental principles of a form of agriculture that respects life and the environment. An agriculture where each farmer must always seek to reconcile ecology, economics and social issues. An agriculture that favours associative nuclei of production through which each farmer will seek to achieve food sëcuritë of his mënage while remaining in solidarity with both the present and future generations. Finally, agriculture practised on areas of amënagëes land recognised and proteëgëes by the public authorities. These guidelines recommend giving priority to:

1° concerted action by farmers, supervisory bodies, local authorities and all those involved in agricultural production in the Commune;

2° the development of farmers by protecting their jobs, their safety and the quality of their human relationships;

3° to make the most of all the food resources that grow around those that are already in use. Making the most of them, as is done with the African square pea (Kikalakasa). And always with future generations in mind. Transforming this production can only extend the duration of this food supply.

4° integrating agricultural activities to save energy ;

5° recognition of these activities and their protection by the public authorities;

6° the preferred use of biofertilisers and biodesinfectants that can be used rationally;

7° quantifying and keeping production statistics,

8° combating all sources of poisoning in fruit, vegetables and meat produced on the premises;

9° Improving species and production ecotechniques and combating all forms of pollution;

10° the permanent availability and accessibility of quality food for the entire population, including the poorest in the Commune.

Let everything that prëcëde be for the food security of households and be conveyed by sustained and continuous mesological education.

Indeed, "there is no ëcodëdevelopment without ëeducation for ecodevelopment". Therefore, environmental education, properly understood, must be global, must extend over the whole of human existence and must reflect the changes of a rapidly transforming universe. I. Environmental education must be open to the community. It must involve individuals in an active process of seeking solutions to society's problems; it must encourage initiative, responsibility and commitment to building a better world. Education, understood in this spirit, is the real catalyst for development. It is the leaven that makes the dough rise, wrote Sachs et al (1981).

In the city of Kinshasa, during the 1990-2000 decade, which was marked by the destruction of the population's economic and health situation, mainly as a result of non-compliance with the resolutions of the Sovereign National Conference, the breakdown of the capital's food supply routes and the effects of the war of liberation waged by AFDI and others, households integrated urban agricultural production into their strategies for combating the serious health problems faced by their members, Households have incorporated urban agricultural production into their strategies for combating the serious health problems faced by their members, and new facts are emerging as new problems that must lead to a new scenario for agriculture adapted to the urban emancipation of Kinshasa 2025-2035. It would also be advisable for urban farmers to adopt new production techniques, including soilless production, permaculture and the improvement of new indigenous or non-indigenous species. Ecological, agronomic and environmental research would be a major asset in strengthening food security.

BIBLIOGRAPHY

Agence Nationale de Mëtëorologie et Tëlëdëtection par Satellite Station de Binza, 2015 Données climatiques de la dëcennie 1997-2006, Kinshasa.

Anonymous, 1984 Memento de l'Agronome 3eme edition Techniques rurales en Afrique, Paris.

Anonymous, 1989 Memento de l'Agronome, 3eme edition Techniques rurales Paris.

Anonymous, 2023 What is the population of the Democratic Republic of Congo in 2023? Demography of the Demographic Republic of Congo. What is the current population of the Demographic Republic of Congo, Population and Demographics Congo Job ecer. com A 106 Welded Fin Tubes Ecer.com official Site & leading Global B2B Platform, published on 25/09/2023.

Ashah G.M.M., 1997 Contraint a l'agriculture urbaine, *Spore* (67), CTA, 10.

Aubert C., 1977 L'agriculture biologique, Pourquoi et comment la pratiquer? Ed. Le Courrier du livre.

Autissier V., 1994 Jardins des villes, jardins des champs-maraichage en Afrique de 1 de Ouest du diagnostic a l'intervention. Collection de Pont. Ed. du GRET, France.

Aya, 2023 Soilless cultivation: advantages and disadvantages Advantages and disadvantages of soil-less cultivation

The disadvantages of soilless cultivation. This is our daily challenge, *AgriMaroc*.Ma AGF SIPCAN Technique.

Azoulay G., 1998 Enjeux de la Securite alimentaire mondiale, *Cahiers Agricultures*, 7(6) CIRAD 1992-2015, 1-7.

World Bank, 2023 What is food security? What is food security and how does the World Bank promote it?

Boulianne M., 2016 Agriculture urbaine, In Anthropen.org, Paris, Editions des archives contemporaines.

Bureau d'Etudes, d'Amenagement et d'Urbanisme- BEAU, 1986 Consommation des produits vivriers a Kinshasa et dans les grandes villes du Zaire, Kinshasa.

BDOM, 1996 Quarterly report on the Kingabwa Health Zone / Hospital Ch. and Liziba Maternity Hospital, Kinshasa.

Bricas, N., 2012 Securite alimentaire in Poulain J.P. (Ed.) Dictionnaire des cultures alimentaires, Paris, PUF.

Brown J.E & Brown R. CMD, 1977 Manuel pour la lutte contre la malnutrition des enfants, un guide pratique au niveau de la Communaute, Kananga Zaire.

CEPLANUT, 1987 Education nutritionnelle -3eme annëe- Besoin nutritionnel des groupes vulnerables, dëpistage de la malnutrition par le perimetre brachial, Kinshasa.

CEPLANUT, 1987 Guide de la fiche de croissance, Saint Paul, Limete Kinshasa.

CEPLANUT, 1994 Etat de la situation nutritionnelle du Zaire, Kinshasa 1994 Kinshasa.

CEPLANUT, 1996 Formation en nutrition pour animateurs et leaders paysans au Kwango-Kwilu, Kinshasa.

CEPLANUT, 1996 Organisation non gouvernementale et securite alimentaire dans la ville de Kinshasa, Kinshasa.

CEPLANUT/FAO, 1994 Etat de la situation nutritionnelle du Zaire, Kinshasa.

CEPLANUT/UNICEF, 1996 Enquete sur la localisation des poches de malnutrition dans la ville de Kinshasa, Kinshasa.

Clements F.E.,1916 Plants succession, an analysis of the development of vegetation Published ST the Carnegie Institution of Washington.

Committee on World Food Security, 2012 Agreeing on terminology, CFS, 39eme session, 5-20 October.

World Commission on Environment and Development, 1988 Our Common Future, Ed. du Fleuve. Les publications de Quebec, Canada.

United Nations Conference on Environment and Development (UNCED), 1993 Agenda 21, Rio Declaration on Environment and Development. Déclaration of Forest Principles, United Nations, New York.

Dajos R., 2000 Precis d'^cologie Dumond, Paris.

Degroote V.A., 1970 Table de composition alimentaire des aliments a l'usage de l'Afrique. FAO Nutrition Paper, Rome, Italy.

Département de l'agriculture, 1987 Programme d'appui et d'organisation du secteur maraicher de Kinshasa, Kinshasa.

Département de l'agriculture et du Développement rural, 1987 Situation actuelle de l'agriculture Zairoise, Kinshasa.

De Rosnay, Sd Le macroscope vers une vision globale, Ed. du seuil Paris.

Dupriey H. & de Leener P., 1987 Jardins et vergers d'Afrique Terres et vie, Ed. L'Harmattan, Apica Enda, CTA, Paris.

Ekofo S., 1989 Introduction aux techniques de tirage des echantillons dans les enquetes socio-économiques, *Cahier Zairois de recherche en Sciences humaines*, Centre de recherche en Sciences humaines, Kinshasa, Republique du Zaire, 1 (1), 163-181.

Essanga T., 1989 L'Enquete par sondage a partir d'un questionnaire applique a la recherche en Education, *Cahier Zairois de recherche en Sciences humaines*. Centre de recherche en Sciences humaines, Kinshasa Republique du Zaire, 1(1), 183-194.

Faburel G., 2003 Vider les villes? Ecologies, le vivant et le social Editions La Découverte coordinated by Philipe Boursier and Cldmence Groumont, Paris.

Favier J.C., Ripert J.I., Laussucq C., Feinberg M. & Ciqual CNVA. 1993 Table de composition des fruits exotiques, fruits de cueillette d'Afrique-Repertoire general des animaux, Tome 3 Ed. ORSOM, TEC DOC, INRA, Paris.

FAO, 1970 Table de composition des aliments a l'usage de l'Afrique, Document sur la nutrition 3, FAO, Rome, Italy.

FAO, 1981 Manuel d'inventaire forestier FAO Forestry Studies 27, FAO, Rome.

FAO, 1964 Les protdines nreud du probldme alimentaire mondial, Italy.

FAO, 2014 Urban agriculture (archive.wikiwix.com " consulted on 3 January 2024.

Giordan A. & Saltet J., 2011 Apprendre a prendre note Librio 999A Ined. Flammario, Paris.

Gobalt J.M., Aragno M. & Matthey W., 2003 Le sol vivant 2[eme] ed. Revue et augmentee Presses Polytechniques et Universitaires Romaines, Lausanne.

Gossens P., Mintem B. & Tollens E., 1994 Nourrir Kinshasa, L'approvisionnement local d'une mdtropole africaine, Ed. L'Harmattan.

Gutu kia Zimi F., 2021 Kinshasa megapole verdoyante en crise. The challenge of urban and environmental management, Author House, USA.

Harlan R.J., 1987 Les plates cultivdes et l'homme, Presses Universitaires de France, Paris.

Houyoux J., 1973 Budgets menagers, nutrition et mode de vie a Kinshasa (Republique du Zaire) Presses Universitaires du Zaire, Kinshasa.

HRP, 2022 Humanitarian Response Plan in DR Congo.

INS, 1984 Totaux definitifs, Zaire, Recensement scientifique de la population Republique du Zaire, Kinshasa.

INS, Departement du Plan, 1989 Enquete budgets menagers Ville de Kinshasa 1985 Principaux resultats. Direction des enquetes economiques, Kinshasa.

INS, 1992 Totaux definitifs, groupement/quartier, vol. 1 Kinshasa, Bas-Zaire, Bandundu, Kinshasa.

INS, 1993 Projection demographique 1984-2000 Zaire et Regions Recensement scientifique de population juillet, Kinshasa.

Information sanitaire au Zaire (ISAZ), 1997 Rapport annuel sur base des données de 1995. Danien M.S.F. Memisa Foundation, Kinshasa.

Kabeya M., Lukebakio N., Kazika K. & Paulus J. Sj., 1994 Inventaire de la flore domestique des parcelles d'habitation Cas de Kinshasa (Zaire), *Revue Medecine et Pharmacopee Africaine*, 8 (1), 58-66.

Kabeya M., Makungu, Mutuba & Paulus J., 1992 Projet " Jardins et Elevage de Parcelle" Rapport annuel 1992, Kinshasa.

Lacoste A.& Salanon R., 1999 Etements de ЫодёодгарbIе et d^cologie 2eme Ed. Nathan Coll. cree par Henri Mitterand, Paris.

Lebrun J. & Stork A. L., 1991 Епитёгайоп des plantes a fleurs d'Afrique tropicale Vol.1 generalite et *Annonaceae* a *Pandaceae*, Conservatoire et jardin botanique de la ville de Geneve, Suisse, Geneve.

Lebrun J. & Stork A. L., 1992 Enumëration des plantes a fleurs d'Afrique tropicale Vol.1I *Chrysobalanaceae* a *apiaceae* avec collaboration de Roger M. Pothil Kew et Delwiens, Utach (*Loranthaceae* et *Viscaceae*, Conservatoire et jardin botanique de la ville de Geneve, Suisse, Geneve.

Lebrun J. & Stork A. L., 1995 Enumëration des plantes a fleurs d'Afrique tropicale Vol.1II Monocotyledones *Limnocharitaceae* a *Poaceae* avec la collaboration de Peter Gold Lott (*Iridaceae*) Conservatoire et jardin botanique de la ville de Geneve, Suisse, Geneve

Lebrun J. & Stork A. L., 1997 Enumëration des plantes a fleurs d'Afrique tropicale Vol.1V *Gamopedales Clethraceae* with the collaboration of Laurant Gautier Geneve (*Sapotaceae*) Conservatoire et jardin botanique de la ville de Geneve, Suisse, Geneve.

Loi fonciere n° 73-021 du 20 juillet 1973, portant Regime Général des Biens, Regime Foncier et Immobilier et Regime des suretes telle que modifiee et completee par la loi n°80- 008 du 18 juillet 1980.

Maisonneuse C.,   1999 La securite alimentaire des menages a Kinshasa. Republique Demographics of the Congo. Action Against Hunger (Action Against-USA) September-December.

Makumbelo C.,   1999 A few developmental possibilities for the improvement of urban agriculture that can contribute to household food sëcurité. Cas de la Commune de Limete, Mëmoire prësentë et dëfendu en vue de l'obtention du grade de Diplome Spëcial (D.S.) en Gestion de l'Environnement Dëpartement de Biologie / Programme de Gestion de l'Environnement, Facu№ des Sciences, Universite de Kinshasa, RD Congo. Kinshasa.

Makumbelo E., Lukoki L., Paulus J. & Luyindula N., 2002 Inventaire des especes vegetales mises en culture dans les parcelles en milieu urbain. Cas de la commune de Limete, Kinshasa, RD Congo, *Tropicultura* 20 (2), 89-95.

Makumbelo E., Paulus J.J.sj., Luyindula N. &, Lukoki L., 2005 Apport des arbres fruitiers a la securite alimentaire en milieu tropical : Cas de la Commune de Limete-Kinshasa

Republique Demographique du Congo, *Tropicultura* 23 (4), 245-252.

Makumbelo E., Lukoki L., Paulus J. & Luyindula N., 2007 Strategie de valorisation des especes ressources des produits non ligneux de la savane des environs de Kinshasa I Enquete ethnobotanique, *Tropicultura* 25(1), 51-65.

Makumbelo E., Lukoki L., Paulus J. & Luyindula N., 2008 Stratëgie de valorisation des especes ressources des produits non ligneux de la savane des environs de Kinshasa II Enquete ethnobotanique, Aspects mëdicinaux, *Tropicultura* 26(3), 129-133.

Makumbelo E., Lukoki L. & Bikoko E., 2016 Enquete ethnobotanique a Kinshasa et carte phytogëographique sur trois especes de *Burseraceae, Revue C.R.I.D. U.P.N.* (066c) Janvier - Mars, 21-29.

Makumbelo C.E., 2023 Gestion durable et strategie d'intervention Cas de la savane de Kinshasa de Wamba (RD Congo), Presses Academiques Francophones.

Makumbelo C.E., Lukoki F.L. & Belesi H.K., 2023 Caracteristiques ecologiques, Evolution et Regeneration naturelle- Reserve et Domaine de Chasse de Bombo Lumene (RD Congo), Editions Universitaires Europeennes.

Mataka K.D., 1994 Protection des ecosystemes et developpement des societes, Etat d'urgence en Afrique, Coll. "Environnement", Ed. L'Harmattan.

Mbemba F. & Remacle J., 1992 Inventaire et composition des aliments du Kwango-Kwilu au Zaire Presses Universitaire de Namur, Belgium.

Mercier J.R., 1980 Energie et agriculture, le choix ecologique, ed. Debard, Paris.

Messiaen C., 1974 Potager tropical-1- Generalite, Coll. techniques vivantes, Presses Universitaires de France, Paris.

Ministry of Agriculture and Community Development, SNV, 1992 Guide de vulgarisation n°1-Culture vivrieres, Kinshasa, SNV/FAO Zaire, April, Kinshasa.

Ministere de l'Agriculture et Developpement communautaire, SNV, 1993 Guide de vulgarisation n°3-Culture maraicheres, Kinshasa, SNV/FAO Zaire, September, Kinshasa.

Ministere de l'Agriculture et Developpement communautaire, SNV, 1994 Guide de vulgarisation n°5-Petits elevages, Kinshasa, SNV/FAO Zaire, April, Kinshasa.

Ministere de l'Agriculture, Animation rurale et Developpement communautaire, 1991 Plan directeur du développement agricole et rural, Kinshasa, May.

Ministries: Plan, Agriculture and Livestock, National Education, Environment, Nature Conservation and Tourism/UNDP/UNOPS, 1998, Securite alimentaire, production et commercialisation, Ville de Kinshasa, Plan triennal (1998-2000) RD Congo, Kinshasa.

Mitja D., 1992 Influence de la culture itinerante sur la vegetation d'une savane humide de Cote d'Ivoire (Booro-Borotou-Touba) Ed. de l'ORSTOM Institut frangais de Recherche scientifique pour le developpement et cooperation Collection Etudes et Theses, Paris.

Monama O., Mukinayi M. & Siku B., 1985 Chaine trophique du plomb, *Revue Zairoise des Sciences nucleaires*, 6 (special), 226-237.

Moran E.F., 1996 Utilisation des connaissances des populations indigenes dans la gestion des ressources de divers ëcosystëmes amazoniens in Alimentation en foret tropicale Interactions bio culturelles et perspectives de dëveloppement Vol. II Bases culturelles des choix alimentaires et stratëgies de dëveloppement l'Homme et biosphere Ed. UNESCO Paris, 1202-1204.

Morard P, 1995 Les cultures vëgëtales hors sol Edition Lavoisier. www. Lavoisier.Fr.

Mosher A.T., 1967 For a modern agriculture. Les impëratifs du dëveloppement et de la modernisation, Coll. Nouveaux Horizons, Les ëd. Internationales, Paris.

Muluma M.G.T., 2003 Le guide du chercheur en Sciences sociales et humaines Les Ed. SOGEDES, RD Congo, Kinshasa.

Musibono D.E., Biey E.M., Kisangala M., Nsimanda C.I., Munzundu B.A., Kekolemba V. &Palus J.J., 2011 Agriculture urbaine comme reponse au chomage a Kinshasa, Rëpublique Dëmocratique du Congo, *Open Edition Journals (VertigO) La revue electroniqueenSciencesdeI' Environnement11* (1). https//:doi.org/10.4000/Vertigo.10818.

Mutuba & Paulus J sj, 1994 Projet "Jardin et Elevage de Parcelle" Rapport annuel, 1994, Kinshasa.

Nations Unies, 1972 Confërences, Environnement et dëveloppement durable, Confërence des Nations Unies sur l'Environnement, du 5 au 16 Juin 1972 Stockholm.

Norma G. & Savard J.G., 1978 Statistiques Les Ed. HRW LEE Canada, Montreal.

Paulus J., Kabeya M., Mutuba N., Musibono E. & Mbemba F., 1989 Role des jardins et ëlevages de parcelle dans 1 alimentation urbaine. Le cas de Kinshasa. Alimentation et nutrition dans les pays en voie de dëveloppement, 4^c ^{me} jouniees scientifiques Internationales du Germ SPA (Belgique) 23-29 avril, Ed. Karthala, ACCT et AUPELF, Paris et Montreal, 45-49.

Pauwels L., 1982 Plantes vasculaires des environs de Kinshasa. Ed. Luc Pauwels, 14, av. G. Vandersmissen, 1040 Bruxelles.

Pauwels L., 1993 N'zayilu Nti, guide des arbres et arbustes de la region de Kinshasa-Brazzaville. Ed. Jardin botanique national de Belgique Meise, Belgium.

Ramade F., 1982 E^ments d^cologie, Ecologie applique, Groupe Mc. Graw Hill Group, Paris.

Sachs I., Bergeret A., Schiray M., Sigal S., Thery D. & Vinaver K.,1981 Initiation a l'Ecodëveloppement, Privat, Toulouse, France.

Srinshw N.S. Taylor G.E. & Gordon, 1991 Interactions between nutritional status and infections WHO, Geneva.

Tazenas du Montcel H., 1985 Le bananier plantain. Ed. Maisonneuse et Larose, Paris.

Touraille C., 1977 Editorial, *Nouvelles de I'Ecodeveloppemenl* (1), CIRAD, France.

International Union for Conservation of Nature and Natural Resources (IUCN), 1980 World Strategy for the Conservation of Living Resources for Sustainable Development. Section 1 Ed. IUCN, UNEP, WWF, Gland Switzerland.

Van Den Abeele M. & Vandenput R., 1956 Les principales cultures du Congo belge 3eme ed. Brussels.

Vandenput R., 1981 Les principales cultures en Afrique centrale, Vandenput R. Editeur, Bruxelles.

Vangu L., 1995 Securite alimentaire au Zaire, contribution de CDI-Bwamanda a la production vivriere (riz, mais, soja, arachide) et l'agriculture familiale 1990-1995 Kinshasa.

White L. & Ann E., 2001Conservation en foret pluviale africaine Methodes de recherche Wide live Conservation Society Premiere Ed. Frangaise, Gabon Libreville.

Wong J.L.G., Thormber K. & Baker N., 2001 13 Assessment of non-wood forest product resources Experience and principles of biometrics Non-wood forest products FAO Italy, Rome.

Wonnacott J.H. & Wonnacott R.J., 1991 Statistique Economique-Gestion -Sciences Medecine (avec Exercices d'application) 4eme Ed. Economia, Paris.

WEBGRAPHY

www.agrimaroc.ma, consulted on 22 December 2023.

www. Congo Job ecer.com, consulted on 5 January 2024.

en.m.wikipedia.org, consulted on 6 January 2024.

www. Lavoisier.Fr, consulted on 22 December 2023.

www.techno-Science.net Urban **ecology-Definition** consulted on 22 December 2023.

www.touphie.org, consulted on 22 December 2023.

www. Un.org, consulted on 10 January 2024.

www.unicef.org>drcongo>children consulted on 25 december 2025.

Table 38. List of food species found at the various survey sites.

Nomen local language	French name (Scientific name)	Types of I'espece	Number of households with this species	Stratum where the species is found
Loso	Marsh rice (*Oryza sativa* L. - *Poaceae*)	Cereale	31	(1)2,7
Matembele	Sweet potato leaf (Ipomoeabatatas (L.)-) *Convulacaceae*)	Lëgume	73	1,2,3,4,5,6,7,9.
Pondu-kautsu	Cassava leaf (*Manihot glaziovii* Miill Arg. - *Euphorbiaceae*)	Lëgume	68	1,2,3,4,5,6,7,9.
Kikalakasa	African square pea (*Psophocarpus scandens* (Engl.) Verdc.- *Fabaceae*)	Lëgume	22	2,4,5,6,7.
Bitekuteku	Amaranth (*AmaranthusviridisL* .- *Amaranthaceae*)	Lëgume	20	1,2,3,4,5,6,7,9.
Ngai-ngai	Common sorrel (*Hibiscus acetocsella* Welw. Ex Hiern - *Malvaceae*)	Lëgume	7	1,6,9.
Ngai-ngai	Sorrel (*HibiscussabdariffaL* .- *Malvaceae*)	Lëgume	7	1,6,9.
Chu	Chinese cabbage (*BrassicaoleraceaL* . var chinensis - *Brassicaceae*)	Lëgume	5	5
Black tip	Blackhead (*Brassica* rapa L. var pekinensis - *Brassicaceae*)	Lëgume	5	1,3,5 .
Tomat	Tomato (*Lycopirsicum cerasiforme* Dunal - *Solanaceae*	Legume	3	6,9.
Dongo dongo	Okra (*Abelmoschusesculentus* (L.) Moench. - *Malvaceae*)	Legume	2	3
Madesu	Bean (*Phaseolus vulgaris* L. - *Fabaceae*)	Legume	2	3,9.
Epinar	Spinach (*Brasella alba* L. - *Basellaceae*)	Legume	2	6
Pilipili	Chilli pepper (*Capsicum frutescens*	Legume	1	6

	L. - *Solanaceae*)			
Mutamwa	- (*Crassocephalumsarcobasis* (DC) S. Moore - *Asteraceae*)	Legume	1	3
Soya	Soya (*Glycinemax* (L.) Merrill-) *Fabaceae*)	Legume	1	9
Kataba	- (JusticiastrataInsularis T.(Anderson) *Acanthaceae*)	Legume	1	3
Minkinka	(*PhytolaccadodecandraL* . I leritier - *Phytolaccaceae*)	Legume	1	9
Bilolo	Morelle (*Solanumaethiopicum-Solanaceae*)	Legume	1	3
Milambwa	Purslane (*Talium triangulare* (Jacq) Willd. *Talinaceae*)	Legume	1	3
Mayebo	Oyster mushroom (*Pleurotus sajor* caju	Mushroom	1	5
Manga	Mango tree (*Mangifera indica* L. (*Anacardiaceae*)	Fruit tree	119	1,2,5,6,7,9.
Avoka	Avocado tree (*Persea americana* Mill. *Lauraceae*	Fruit tree	87	1,2,5,6,7,9.
Payi payi	Papaya (*Carica papaya* L. *Caricaceae*)	Fruit tree	59	1,2,3,4,5,6,7,9
Mbila	Oil palm (*Elaeis guineensis* Jacq *Arecaceae*)	Fruit tree	59	1,2,5,6,7,9.
Makemba	Banana (*Musa paradisiaca* L. Musaceae	Fruit tree	59	1,2,3,4,5,7,9.
Nsafu	Safoutier *Dacryodes edulis* (G.Don) H.J. Lam *Burseraceae*)	Fruit tree	30	1,2,3,5,6,7,9.
Mundenge	Creur de breuf (*Annona squamosa* L. *Annonaceae*)	Fruit tree	21	1,2,3,5,6,7,9.
Sitro	Lemon tree (*Citrus limon* (L) Burm.f. Rubiaceae	Fruit tree	21	2,3,6.
Pom	Apple tree	Fruit tree	19	1,2,3,4,5,6,7,9.

	(*Eugenia malaccensis* L. *Myrtaceae*)			
Madarine	Madarinier *Citrus reticulata* Blanco *Rutaceae*	Fruit tree	15	2,3,6.
Koko	Sugar cane (*Saccharum cofficinarum* L. *Poaceae*)	Fruit tree	11	1,2,2,5,6,7,9.
Kofitur	Prumier de (*Flacourtia ramontchi* L. Merit *Flacourtiaceae*)	Fruit tree	6	2,5,9.
Madamë	Badamier *Terminalia catappa* L. *Combretaceae*)	Fruit tree	6	1,5,6,7,9.
Nzeteya santu Petelo	Bread tree *Artocarpus altilis* (Park.) Fosberg *(syn. Artocarpus incisa* L.f.) Moraceae	Fruit tree	5	2,5,6,9.
Guava	Guava (*Psidium guajava* L. (Myrtaceae)	Fruit tree	5	1,5,6,9.
Pakapaka	Carambolier (*Averrhoa carambola* L. (*Oxalidaceae*)	Fruit tree	1	1
Soso	Hen (*Gallus Gallus* (*Gallus domestica*) *Gallinacees*	Animal species	43	1,2,3,5,6,7,9.
Libata	Common duck -Baribarie (*Anas platyrineros Austedias*)	Species animal	13	2,3,5,6,7,9.
Pigeon	Pigeon (*Collumba livia Collumbidae*)	Animal species	6	1,5,6.
Kobaye	Guinea pig (*Covia porcellus* L. *Cavidae*)	Animal species	2	5,6.
Ngulu	Pork (*Sus domesticus* *Sus sera fa domesticus Suidae*)	Animal species	2	5,6.
Taba	Goat (*Carra Hircus domestica* *Caprides*	Animal species	1	6
Kanga	Common guinea fowl (*Numida* *mibagris* *Phaesiamides*)	Animal species	1	6

I want morebooks!

Buy your books fast and straightforward online - at one of world's fastest growing online book stores! Environmentally sound due to Print-on-Demand technologies.

Buy your books online at
www.morebooks.shop

Kaufen Sie Ihre Bücher schnell und unkompliziert online – auf einer der am schnellsten wachsenden Buchhandelsplattformen weltweit! Dank Print-On-Demand umwelt- und ressourcenschonend produzi ert.

Bücher schneller online kaufen
www.morebooks.shop

info@omniscriptum.com
www.omniscriptum.com

Printed by Books on Demand GmbH, Norderstedt / Germany